Boundary Layer Flows - Advances in Experimentation, Modelling and Simulation

Edited by Zambri Harun, Andrea Aprovitola and Giuseppe Pezzella

Published in London, United Kingdom

Boundary Layer Flows - Advances in Experimentation, Modelling and Simulation
http://dx.doi.org/10.5772/intechopen.111067
Edited by Zambri Harun, Andrea Aprovitola and Giuseppe Pezzella

Contributors
Ashraf Amer Abbas, David Weyburne, Dongming Chen, Gan Tian, Haifeng Zhang, J.A. Virbalis, Mattias K. Gustavsson, R. Račkienė, Siddharth Rout, Wenjun Yuan, Zambri Harun

First published in London, United Kingdom, 2024 by IntechOpen
IntechOpen is the global imprint of INTECHOPEN LIMITED, registered in England and Wales, registration number: 11086078, 167-169 Great Portland Street, London, W1W 5PF, United Kingdom

British Library Cataloguing-in-Publication Data
A catalogue record for this book is available from the British Library

Additional hard and PDF copies can be obtained from orders@intechopen.com

Boundary Layer Flows - Advances in Experimentation, Modelling and Simulation
Edited by Zambri Harun, Andrea Aprovitola and Giuseppe Pezzella
p. cm.
Print ISBN 978-1-83769-835-6
Online ISBN 978-1-83769-834-9
eBook (PDF) ISBN 978-1-83769-836-3

For EU product safety concerns:
IN TECH d.o.o., Prolaz Marije Krucifikse Kozulić 3, 51000 Rijeka, Croatia,
info@intechopen.com or visit our website at intechopen.com.

Meet the editors

Dr. Zambri Harun is a senior fellow at the Faculty of Engineering and Built Environment, National University Malaysia (UKM). He graduated with a BSc and MEng, both from Rensselaer Polytechnic Institute, New York, USA. His experience in the industry stretches over almost a decade; being an engineer at Motorola Malaysia and in a civil project. His Ph.D. was on turbulence flow from the University of Melbourne, Australia. His research covers wind tunnel and atmospheric turbulence as well as building and tunnel flow management. He has published more than 150 peer-reviewed articles in the areas of wall-bounded flows, wind energy, urban heat island (UHI), emission patterns, and end-of-life vehicle (ELV) initiatives.

Dr. Andrea Aprovitola is a research fellow at the Department of Engineering, University of Campania, "L. Vanvitelli," Italy. He is currently involved in reentry vehicle design activities, specifically focusing on fluid dynamic studies and the design of innovative winged-body aerodynamic configurations for space reentry. His work includes simulations using finite-volume methodologies for both low and high Mach number turbulent flows during reentry, as well as new blended wing-body configurations. Additionally, he is developing and implementing multidisciplinary analysis procedures to optimize the aerodynamic performance of reentry configurations. His research activity is also focused on the calculation of thermo-aerodynamic characteristics of an aircraft usable for atmospheric reentry (lifting body). Currently, such evaluations are carried out with the open-source code SU2 and compared with experimental data as well as results obtained with established commercial codes. He is also experienced in turbulence modeling with RANS and LES methodologies, in high-order finite volume discretization.

Dr. Giuseppe Pezzella is a lecturer in Aircraft Aerodynamics and Aerothermodynamics at the Department of Engineering, University of Campania, "L. Vanvitelli," Italy. He was also a senior research engineer at CIRA (Italian Aerospace Research Centre) in the field of hypersonic aerodynamics and aerothermodynamics. He obtained an MSc in Aerospace Engineering in 1999 and a Ph.D. in Aerospace Engineering in 2004 at the University of Naples "Federico II," Italy. He is currently involved in reentry vehicle design activities in both international and national research programs.

Contents

Preface **IX**

Section 1
The Structure of the Boundary Layer **1**

Chapter 1 **3**
Wind Tunnel Measurement Techniques
by Zambri Harun and Ashraf Amer Abbas

Chapter 2 **23**
Novel Approach for Turbulent Flow- and Onset Analysis
by Mattias K. Gustavsson

Section 2
Drag Reduction Initiatives Using Active Control **47**

Chapter 3 **49**
On Drag Reduction by Spanwise Wall Oscillation in Compressible Turbulent Channel Flow
by Wenjun Yuan, Dongming Chen and Haifeng Zhang

Chapter 4 **65**
Active Flow Control and Its Applications in Supersonic Boundary Layer
by Gan Tian

Section 3
Verification and Applications of Flow Mechanics **85**

Chapter 5 **87**
Falkner-Skan Experimental Verification Failure for Airflow on a Wedge
by David Weyburne

Chapter 6 **99**
The Influence of Flow Admixtures to the Electromagnetic Flow Meter Accuracy
by R. Račkienė and J.A. Virbalis

Chapter 7 **123**
Early Advancements in Turbulence-Generated Noise Modelling: A Review
by Siddharth Rout

Preface

Boundary Layer Flows – Advances in Experimentation, Modelling and Simulation is organized into three sections.

The first section, "The Structure of the Boundary Layer", highlights boundary layer measurement techniques using the classic method of hotwire anemometry.

A wind tunnel in favorable pressure gradient (FPG) settings, with pressure gradient parameters $\beta = -0.3$ to -0.5 in boundary layer thickness Reynolds number (Re_τ) ranging from 1200 to 3400, was compared between two facilities: the Pangkor Low-Speed Wind Tunnel (PLSWT) at Universiti Kebangsaan Malaysia (UKM) and the Green Tunnel at the University of Melbourne, Australia. Using the Clauser chart method to collapse the velocity profile revealed the same near-wall and logarithmic layer but an increasing wake region with Re_τ. Turbulence intensities and energy spectra show that turbulence intensities peak at $z^+ = 15$, with the signature near-wall features having a length scale $\lambda_x^+ = 1000$, both remaining unchanged regardless of Re_τ.

A computation of 20 different turbulent flow cases of fully developed pipe flows, for airflow at 20°C in a 14-cm diameter tube with varying surface roughness, was used to develop a novel approach to modeling turbulent flows. Turbulence transition can be accurately predicted at pipe diameter Reynolds number Re_D around 2300 for a fully developed pipe flow using this model.

The second section, "Drag Reduction Initiatives Using Active Control", discusses high-speed drag reduction initiatives, which are attracting more attention as modernization requires more high-speed traveling means. The active control mechanism is just one of the methods to achieve drag reductions by way of manipulating the most crucial parts of the boundary layer. In the first experiment, a supersonic boundary layer with a speed of Mach 2.0 to 3.0 is studied with visualization. The different methods of more efficient boundary layer transition were used including the use of plasma-based actuators as turbulators to increase the turbulence. Surface arc plasma actuator (SAPA) arrays were used successfully as an active control in an experiment to perform boundary layer transition on a flat plate.

Next, the drag reduction initiatives using direct numerical simulations (DNS) reveal a reduction of 39.3% for supersonic compressible turbulent channel flows. The reduction is initiated by the spanwise motion generated by the wall movement. It is found that the reduced drag reduction is mainly due to the decreased turbulent contribution under the action of oscillatory walls. Dealing with supersonic flows at bulk Mach number $M_b = 1.5$ is challenging, as the CPU facilities need to be as large as 5×10^5 CPU core hours-equivalent to obtain statistically stable results.

In the last section, "Verification and Applications of Flow Mechanics", the book highlights the verification and applications of flow mechanics. Not all fluid mechanics results are repetitive. A Falkner-Skan flow experimentation to a boundary layer flow could not be extensively verified using an OpenFOAM package. The exception that did show good profile similarity for both velocities and the pressure gradient, the Blasius flat plate flow model, suggests the need for more study into wedge-type boundary layer development where achieving the intended pressure distribution is very challenging.

Applications of flow mechanics in measuring devices are crucial for ensuring accurate flow calculation. One area of focus is electromagnetic flow meters, especially when the flow contains admixtures with magnetic and/or electric properties different from the fluid. Studies on fluid with varying admixture content were conducted, resulting in a general expression of measurement error for common cases.

Noise generated from turbulence is a big issue, akin to noise generated by tangible devices such as fans and motors. The book discusses groundbreaking research on quantifying noise generation from conservation laws, including the Navier–Stokes equations, and its use in approximating acoustic intensity through acoustic analogy. Progress in understanding noise generation has been applauded, as it facilitates the development of effective mitigation plans.

Zambri Harun
Department of Mechanical and Manufacturing Engineering,
Faculty of Engineering and Built Environment,
National University of Malaysia,
Bangi, Malaysia

Andrea Aprovitola and Giuseppe Pezzella
University of Campania "Luigi Vanvitelli",
Italy

Section 1

The Structure of the Boundary Layer

Chapter 1

Wind Tunnel Measurement Techniques

Zambri Harun and Ashraf Amer Abbas

Abstract

This chapter discusses different measurement techniques of a low-speed wind tunnel designed and built at the Faculty of Engineering and Built Environment, Universiti Kebangsaan Malaysia (UKM). The fully automated wind tunnel is named Pangkor after an island in Perak, Malaysia. Different measurement techniques are used to understand and validate the flow quality of the turbulent boundary-layer profiles at different locations and directions (wall-normal and spanwise). A hot-wire sensor captures the boundary layer over a smooth, flat surface. These data are then compared with high-quality published data. The flow uniformity shows comparable velocity variations and turbulent intensity. The boundary-layer profiles collapse well in different spanwise locations. Furthermore, the boundary-layer profiles at different Re_τ follow the standard boundary-layer profile. Since the wind tunnel is relatively new, the calibration method for the hot-wire anemometry is provided. The approach to spectral analysis is provided using a fast Fourier transform (FFT), revealing the prominence of the energized structure of $\lambda_x{}^+ \approx 1000$ residing in the near-wall, in this analysis, we chose $z^+ \approx 15$. The spectral analysis shows that the compounded effects of small and large-scale do not vary with Re_τ. Nonetheless, the outer hump is increasingly visible with increasing Re_τ. The proposed measurement technique and findings help validate wind tunnel flow quality and turbulent boundary layer profiles.

Keywords: wind tunnel, turbulent boundary layer, turbulent flow, hot-wire anemometer, boundary layer experiments

1. Introduction

Wind tunnels have been essential for aerodynamic research since the dawn of aviation [1]. Despite changes in their role and occasional predictions of their obsolescence, this situation is unlikely to change in the foreseeable future. Therefore, up-to-date measurement techniques are worthwhile. Constructing a new low-speed wind tunnel to study the fundamentals of turbulent boundary layers is a challenging endeavor that requires a profound comprehension of boundary layer dynamics and wind tunnel design [2]. These tunnels must undergo thorough testing to simulate typical boundary layer profiles with minimal turbulence intensity accurately. Extended wind tunnel studies are essential for advancing novel aircraft, wind turbines, and other systems that delicately interact with airflow. Low-speed aerodynamics is gaining popularity in both industry and academics. Despite the progress made in numerical models, low-speed wind tunnels remain essential in research and design procedures due to their unparalleled capacity to offer precise real-world data [3]. A smooth, turbulent boundary layer is necessary for sensitive experiments, where a high-quality flow becomes increasingly important [4]. One of the main variables affecting flow in a wind tunnel is the severity of turbulence. During the wind tunnel design process, much work is put into ensuring that the test section flow may have low levels of turbulence and stability. The main methods employed in the flow straightening and turbulence reduction system are contractions, honeycomb, and wire-mesh screens [5]. Most studies agreed that the screens and contractions reduce turbulence and mean velocity variation more in the longitudinal than in the lateral direction [3]. In the present research, a new low-speed wind tunnel was built in the Mechanical and Manufacturing Engineering Department, Universiti Kebangsaan Malaysia. This study aims to provide different measurement techniques using hot-wire anemometer. This method would help understand the turbulent boundary layer flow in wind tunnels.

2. Methodology

This study intends to systematically evaluate and validate the turbulent boundary layer profiles in a newly built wind tunnel. An extensive amount of data was gathered by employing various measuring approaches, each offering a unique set of benefits. A complete comparison and validation of the characteristics of the turbulent boundary layer were carried out by comparing the current data to high-quality published data [6, 7]. This was accomplished by employing sophisticated analytical tools to evaluate the data. A more reliable and in-depth understanding of turbulent boundary layers is vital for advancing wind tunnel and fluid mechanics.

2.1 Wind tunnel layout

Despite being owned by the Mechanical and Manufacturing Engineering Department, the Pangkor low-speed wind tunnel (PLSWT) is located at the Coastal and Water Resources Laboratory of the Faculty of Engineering and Built Environment, Universiti Kebangsaan Malaysia (UKM). The laboratory belongs to the Department of Civil Engineering, where research on waves, hydraulics using a wave maker, and channel flows are conducted. The wind tunnel has been developed and fine-tuned for the current research to improve flow quality and the turbulent boundary layer pattern. The wind tunnel is powered by a 15 kW AC axial fan, followed by a flexible duct, transition duct, diffuser, settling chamber, contraction nozzle with an area ratio of 2.4:1 and a test section with cross-sectional area 1.2 × 0.476 m (width × height) and 3 m total length. **Figure 1** shows a general outline of the wind tunnel's main parts. **Figure 2** shows the actual photo of the wind tunnel.

The design and construction of the PLSWT started in 2014. The wind tunnel operated for the first time in September 2015 with a turbulence intensity level of 5%. To start with, the wind tunnel had only minimal measures to address turbulence intensities, that is, two meshing screens at the diffuser section. However, this value needed to be more suitable for the researchers to perform experiments to study the fundamentals of the turbulent boundary layer. In 2017, the second stage started with combining honeycomb and meshing screens to fine-tune the turbulent

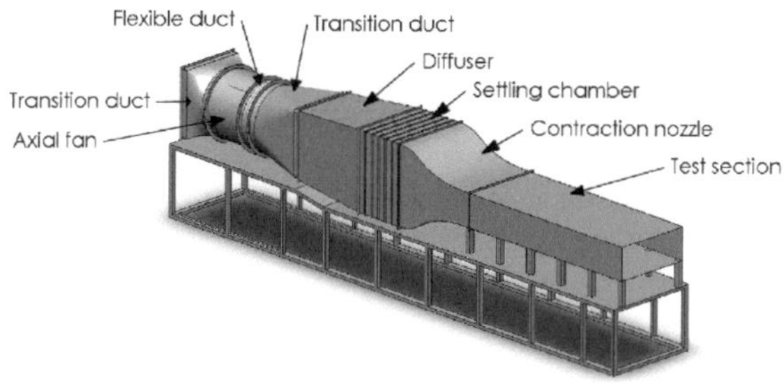

Figure 1.
General outline of Pangkor low-speed wind tunnel.

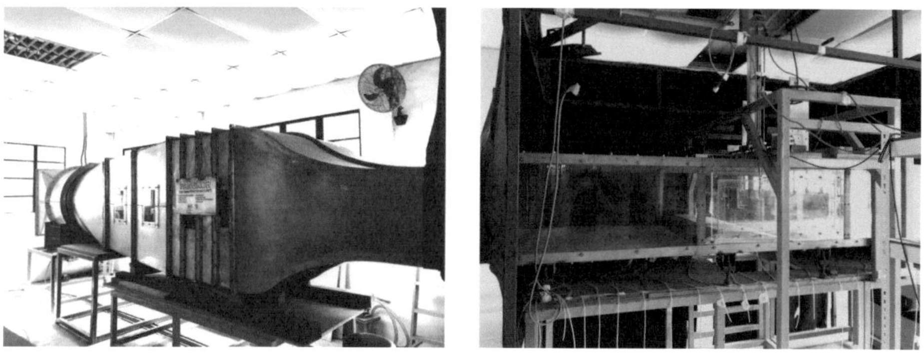

Figure 2.
Pangkor low-speed wind tunnel. The left photo shows the main parts of the wind tunnel, starting from the transition duct to the contraction nozzle. The second photo shows the test section, and the traverse system is on the top of the ceiling.

boundary layer profile and flow quality. Four new layers of meshing screens and a low-cost honeycomb built in-house, a three-dimensional plywood frame with dimensions of height, h = 1200 mm, width, w = 1200 mm, length, l = 200 mm, and thickness, t = 3 mm, were used as the main frame structure for the honeycomb. This frame is filled with 10,000 cells of circular plastic straw with diameter dimensions $d_s = 12$ mm, length $l_s = 177.5$ mm, and straw thickness $t_s = 0.2$ mm. The honeycomb was installed in the middle of the setting chamber, where two layers of meshing screens were placed downstream of the honeycomb device. The last two layers are installed upstream to the honeycomb wall, to be six layers of the meshing screen in total [5]. For further information about the calculations of the meshing screen and honeycomb design, see Refs. [5, 8, 9]. The primary purpose of the honeycomb device and meshing screen in the wind tunnel is to support the flow homogeneity and the turbulence fluctuation and also to adjust the direction of the incoming flow to the test section. They resulted in a very low free-stream turbulence level of 0.0085% (see **Figure 3**).

In early 2018, serious development and fine-tuning for the wind tunnel were achieved. A new floor made of solid wood has been installed. Also, three stages of tripwires were used to develop the boundary layer. Therefore, this converted the wind tunnel to be the first for turbulent boundary layer studies in Malaysia with a minimum velocity variation and turbulence intensity value. For further information about the list of wind tunnels in Malaysia, see a review paper by Wiriadidjaja et al. [10].

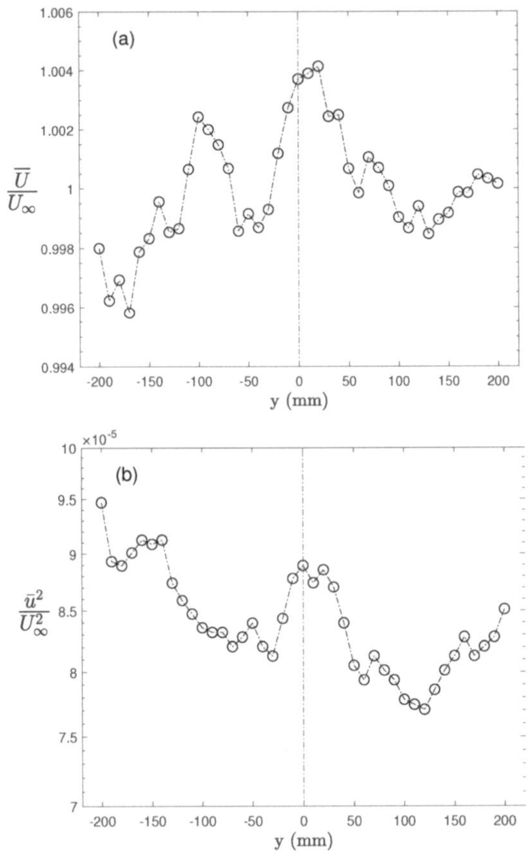

Figure 3.
The variation of the mean velocity profile (a) and turbulence intensity (b) over the spanwise direction (y) of the wind tunnel. The vertical red dashed line indicates the center of the test section in a vertical direction.

2.2 Traverse construction

The traversing mechanism regulates the motion required for the installed sensors, such as the Pitot tube, hot-wire, or temperature probes, in both the vertical and horizontal directions. This enables various motions that may cover many locations within the examined sample. A two-dimensional (2D) traverse system designed and manufactured in-house includes two sliders firmly positioned within two sets of guides. These guides can be positioned vertically or horizontally and sit above the ceiling of the test section. Two fine-stepper motors facilitate the propulsion for this traverse. Therefore, each revolution necessitates 800 steps. The Vecta Steppers motors, specifically the PK266-034 model, are connected to Velmex controllers, the VXM-3 model. These motors are fitted with ball screws with a pitch of 1.6 mm, which applies to vertical and horizontal movements. The traverse provides a level of precision of less than 0.1 mm, or 4×10^{-3} mm for each step. This precision is necessary to capture the velocity fluctuations especially in the near-wall region. The motor controllers provide good support because communication can be established with many programming languages.

2.3 Data acquisition and automation

The most essential features of the wind tunnel are its functionalities and ease of use. A single-point control is preferred because operations can be made simple. Scilab is the only software for controlling air flow speeds, traverse systems, and acquisition operations. Starting from the motor, a National Instrument (NI) data acquisition (DAQ) system sends signals to the variable frequency drives (VFD). The fan speed was controlled through an output voltage signal produced by the NI DAQ 6008 module. This procedure is essential for hot-wire calibration. The atmospheric conditions, for example, atmospheric pressure (p), temperature (T), and relative humidity (RH), were measured using the Comet H7331 device for all experiments. The device was connected to the PC *via* a USB connection. The Modbus protocol was used to transfer data from the apparatus to Scilab. A simple Scilab code can then be elaborated to get the data from Comet H7331. A Velocicalc-TSI/9565A device was installed to get the mean flow velocity. The instantaneous velocity was measured using a Dantec hot-wire anemometer. A Dantec adapter conditioned the signal of the hot wire and then an NI DAQ 9215 module, which is connected to the Multichannel Constant Temperature Anemometers (CTA) and, therefore, *via* USB connected to the PC. Data were collected using an existing data acquisition box connected to a laptop using Scilab. The current data acquisition box can handle high requirements of at least a 100 kHz sampling rate with at least four channels in operations. A high-frequency rate is required to detect the energetic near-wall and large-scale features using the energy spectrum method [5, 11].

2.4 Calibration procedure

The hot-wire sensor is calibrated within the wind tunnel in all experimental series, employing a Pitot tube. The calibration process is conducted twice throughout each experiment, known as pre-calibration and post-calibration. To properly calibrate the hot-wire sensor, it is necessary to position it at a free-stream location, specifically at the center of the test section. Following that, it is essential to systematically measure and record a series of velocities, incrementally ranging from zero up to about 50% beyond the free-stream velocity deemed necessary for the experiment. Consequently, a polynomial third-order fitting is employed to establish a relationship between the voltage measured by the hot wire and the velocity measured by the Pitot tube. A comparison sample of calibration data is presented in **Figure 4**. This example includes both pre-calibration and post-calibration data. The pre-calibration and post-calibration curves are not expected to collapse because of expected temperature drift. A temperature drift of 1°C is usually acceptable.

Even though there is an increasing variation as velocity is increased as shown in **Figure 4**, the current comparison is deemed acceptable. When there is a large temperature drift that can be caused by the wind tunnel fan that keeps producing heat or simply because the building absorbs more energy as the sun rises, a temperature compensation procedure can be implemented. A temperature compensation is a procedure when an interpolation between pre- and post-calibration is performed. A typical acceptance criterion is the free-stream velocity should be within 2%.

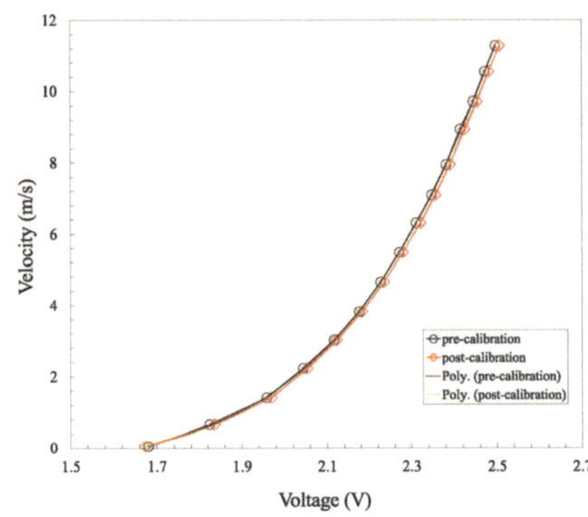

Figure 4.
Hot-wire calibration curve for $U_\infty \approx 12$ m/s; where ∘ is pre-calibration and ⋄ is post. Poly. stands for polynomial fitting of the data.

2.5 Determining initial distance between wall and hot wire

A manual measurement established the distance between the hot-wire sensor and the wall. The distance was determined accurately by employing a high-resolution camera and a ruler. An exact traverse system was crucial for effectively controlling the exact movement of the hot wire in millimeter increments. Consequently, the hot-wire sensor was positioned successfully at approximately z = 0.25 mm from the wall. This distance directly influences the effectiveness of capturing the inner region of the turbulent boundary layer, particularly the inner peak. The positioning of the hot wire, including its inclination and yaw angle, followed the proposed method by Hutchins & Choi [12]. It is essential to use the strategy that has been provided in order to traverse the hot wire close to the wall before prong contact takes place. However, the traverse movement is no longer immediately connected to a wire movement after any wall contact has been made. The consequence at this point is that it leads to inaccurate measurements or hot-wire deformation (broken hot-wire sensor) (**Figure 5**).

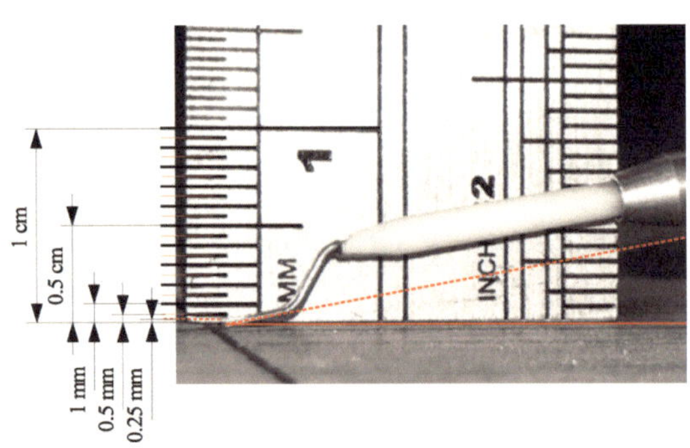

Figure 5.
Determining the distance between the wall and hot wire using the physical method.

3. Results and discussion

3.1 Flow uniformity

In order to ensure that the wind tunnel has a high-quality flow condition, a spanwise measurement was performed to analyze the velocity and turbulence intensity variability profiles. The measurements were conducted at the center of the test section at a distance of wall-normal z = 250 mm and streamwise distance of x = 1.5 m with a free-stream velocity of $U_\infty = 1.6$ m/s. The velocity and turbulence intensity along the center line of the wind tunnel are shown in **Figure 3**. A hot-wire sensor was used to measure the velocity fluctuation at a frequency of 50 kHz for 180 seconds to ensure sufficient data capture. In the current measurements, the hot-wire platinum wire of 5 μm was used with a length of approximately 1.5 mm, and an overheat ratio was set between 1 and 1.5, similar to typical boundary layer studies [11, 13]. Unfortunately, the traverse movement is limited to only 400 mm in the spanwise direction (y), resulting in a total of 41 measurement points with a separation of 10 mm between each point. The velocity variation profile $\overline{U}/U_\infty$ plotted as a function of the spanwise direction (y) is shown in **Figure 3(a)**, where the maximum deviation from the velocity profile $\approx$ 0.8%. The scaled turbulent intensity profiles $\overline{u}^2/U_\infty^2$ are shown in **Figure 3(b)**. The profile indicates that the turbulent intensity variability is less than 10%, which is considered within acceptable limits. The mean value of the turbulent intensity in the wind tunnel $\approx$ 0.0085%. In summary, the wind tunnel exhibits good flow conditions to perform fundamental investigations on turbulent boundary layers and engineering applications, including airfoils, wind turbines, and surface roughness.

3.2 Turbulent boundary layer variability

A further check on the flow condition was performed. However, the focus this time is not on the homogeneity of flow in the center of the test section but on evaluating the quality of the turbulent boundary layer profile in various spanwise directions (y), as indicated in **Figure 6**. This is done to guarantee that the wind tunnel's boundary layer profile follows the standard pattern and does not suffer from

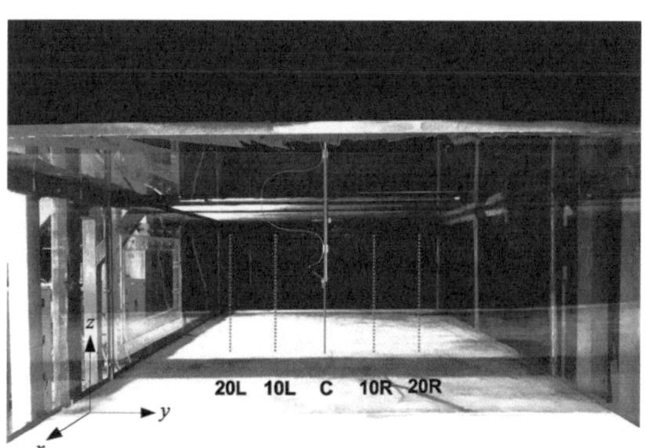

Figure 6.
Schematic diagram of the turbulent boundary layer measurements in different spanwise locations (y). The solid red line indicates the center of the test section, denoted by the letter C.

any defect. **Figure 6** shows the wind tunnel test section and the location of spanwise measurements. The solid red line indicates the center of the test section, denoted by the letter C. The dashed red lines show the spanwise location of the turbulent boundary layer measurements denoted by 20 L, 10 L, 10R, and 20R. The letters L and R indicate the left and right directions about the center. The meaning of 10 L is that the measurement point shifted 100 mm to the left direction from the center point (C), and 20 L is shifted 200 mm from the center point (C). The points 10R and 20R shifted in the right direction from the center point (C). In the current experiment, there are five measurement points, each point separated by 100 mm, thus a coverage of 400 mm.

The wind tunnel is set to a free-stream velocity of $U_\infty \approx 5\,\mathrm{m/s}$ subjected to favorable pressure gradient (FPG) environment. Measurements were made with a locally fabricated sensor soldered to a single hot-wire boundary layer type prob. with a core diameter of 5 microns (μm) [14]. All measurements were performed at the same streamwise distance $x/l = 50\%$ in the test section length percentage, where l is the test section length. In each boundary layer measurement, the hot-wire sensor traverses 50 logarithmically spaced wall-normal positions starting at approximately 0.250 up to 200 mm. All measurements were performed at a frequency of 20 kHz for 180 s, which represents boundary layer turnover times of $TU_1/d \approx 94000$ (T is time in seconds, U_1 is local free-stream velocity, and δ is the boundary layer thickness), which is more than sufficient for capturing the large-scale features in the near-wall region [7, 15, 16]. This section discusses and compares the statistical analysis of all five points over the smooth surface of the wind tunnel by analyzing the diagnostic plot, mean velocity, velocity defect, and turbulent intensity.

3.2.1 Diagnostic plot

The diagnostic plot was introduced by Alfredsson et al. [17] and Alfredsson et al. [18] as a method for evaluating the quality of the turbulent boundary layer. One significant feature of the diagnostic plot is that it does not require knowing the wall position or the friction velocity, therefore ensuring the elimination of any confusion relating to the position of the wall and the velocity related to friction. In their investigation of a smooth-wall turbulent boundary layer subjected to a zero pressure gradient (ZPG) environment [17–20], it has demonstrated that the turbulent intensity, represented by u_{rms}/U, exhibits a linear decline with U/U_∞ throughout the entire range spanning from the logarithmic to the wake region. Moreover, it can be shown that the linear zone expands as the friction Reynolds number ($Re\tau$) grows. In this region, all profiles exhibit a satisfactory collapse within the range of $0.5 \le U/U_\infty \le 1$. **Figure 7** shows the diagnostic plot over the smooth surface of the wind tunnel bed for five different locations in the spanwise direction. All profiles show a clear pattern and collapse very well for the entire region within the range of $0.6 \le U/U_\infty \le 1$, which indicates that the turbulent boundary layer profiles in the current wind tunnel do not experience any defects and have high-quality data.

3.2.2 Turbulence statistics: velocities and turbulence intensity profiles

Given the lack of direct measurements of τ_w, the frictional velocity U_τ of the smooth wall needs to be estimated from the measured velocity profiles. The current

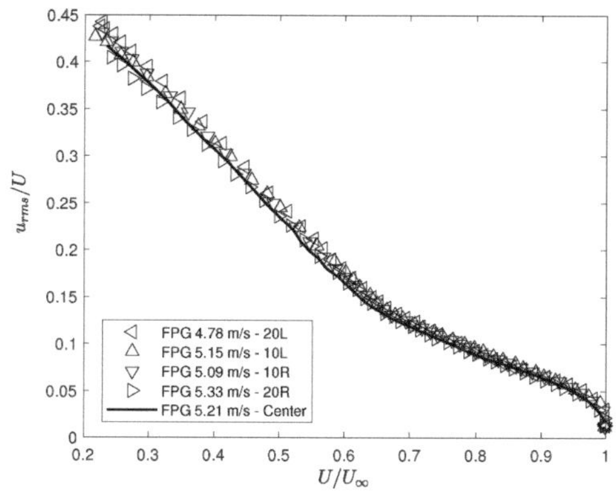

Figure 7.
Diagnostic plot of smooth turbulent boundary layer variability. Symbols (◁) represent the point at 20 left, (Δ) 10 left, (∇) 10 right, (▷) 20 right, and the solid black line shows the measurements at the center of the test section.

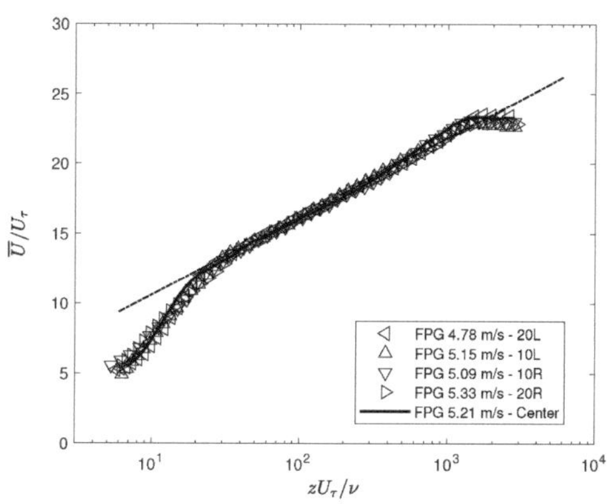

Figure 8.
Comparison of the mean velocity profiles of turbulent boundary layer of five points in the spanwise direction (y). The dashed-dot black line indicates the log law, here κ = 0.41 and a = 5.0. Symbols (◁) represent the point at 20 left, (Δ) 10 left, (∇) 10 right, (▷) 20 right and the solid black line shows the measurements at the center of the test section.

study utilizes the Clauser method [21, 22] to determine the value of U_τ. This is achieved by fitting the measured mean velocity to a log-law function with constants $kappa = 0.41$ and an intercept A = 5.0 [23]:

$$U^+ = \frac{1}{\kappa} \ln\left(\frac{zU_\tau}{\nu}\right) + A, \tag{1}$$

Figure 8 shows the mean velocities profile comparison between five locations in the spanwise direction. The velocity scaled with the skin-friction velocity $U\tau$ and the wall-normal length is made nondimensional by inner scaling of ν/U_τ, where ν is the kinematic viscosity. It shows a good agreement between all five profiles and collapses well in the near-wall region, logarithmic later, and the wake region.

Figure 9 shows the velocity defect of all five profiles. The experiment data show an excellent agreement for the entire boundary layer. However, the current profiles do not follow the standard behavior of a turbulent boundary layer subjected to zero pressure gradient (ZPG). When the boundary layer is subjected to ZPG environment,

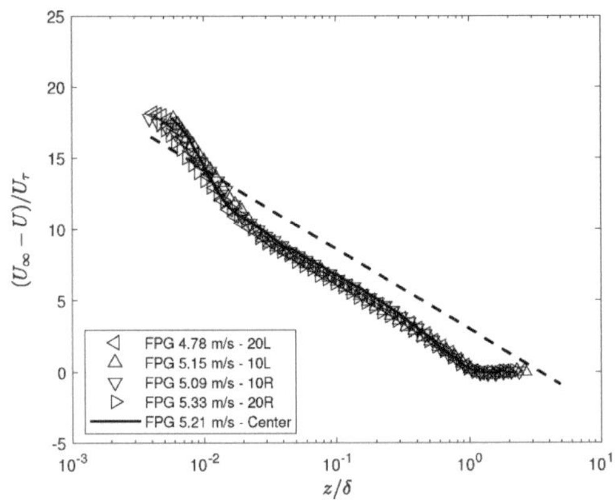

Figure 9.
Comparison of the velocity defect of five points in the spanwise direction. The dashed black line represents the defect low. Symbols (◁) represent the point at 20 left, (Δ) 10 left, (∇) 10 right, (▷) 20 right, and the solid black line shows the measurements at the center of the test section.

the outer layer similarity profiles collapse well and agree with Townsend's outer layer similarity [7, 24]. In other words, all the profiles in the ZPG case collapse well with the velocity defect low, represented by the following equation,

$$\frac{U_\infty - U}{U_\tau} = -\frac{1}{\kappa}\ln\left(\frac{z}{\delta}\right) + B \tag{2}$$

where κ is the Karman constant, and B is the velocity defect constant; in the current study, B = 3. In the current study, all profiles collapse well. However, they shifted downward due to the effects of a favorable pressure gradient (FPG). The pressure gradient data provided by Harun [7] showed effects similar to those of the current experiment when the boundary layer was subjected to a favorable pressure gradient (FPG). The opposite effects can be seen when the boundary layer was exposed to an adverse pressure gradient (APG), where the velocity defect profile is

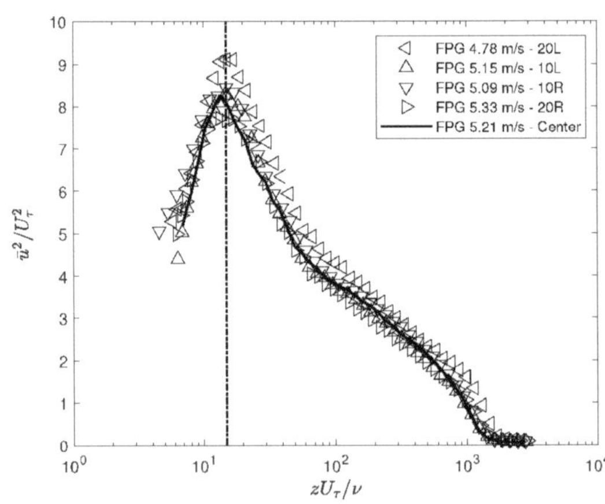

Figure 10.
Comparison of the turbulent intensities of five points in the spanwise direction. Symbols (◁) represent the point at 20 left, (Δ) 10 left, (∇) 10 right, (▷) 20 right, and the solid black line shows the measurements at the center of the test section. The black dashed-dot line indicates the location of inner peak at z^+ = 15.

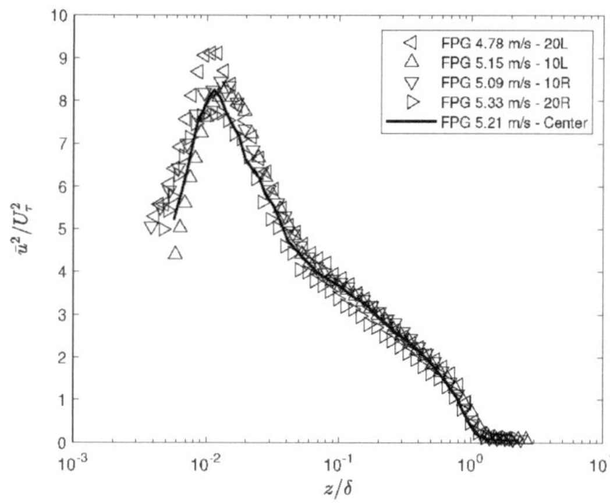

Figure 11.
Comparison of the turbulent intensities of five points in the spanwise direction. Symbols (◁) represent the point at 20 left, (Δ) 10 left, (∇) 10 right, (▷) 20 right, and the solid black line shows the measurements at the center of the test section.

shifted upward. The downward and upward shifts are due to the difference in skin friction velocity between the FPG and APG environments.

Figure 10 compares the turbulent intensity profiles of all five locations. The inner peak follows the standard boundary layer profile at $z^+ = 15$, which has the highest level of turbulent intensity. There is an entirely different level of turbulent production in the near-wall region due to a lower free-stream velocity, particularly at the location 20 L, compared to other experiments. This results in a lower skin-friction Reynolds number Re_τ, therefore, low frictional velocity, which affects the scaling of the profiles. However, the differences between the profiles on the near-wall region are acceptable. In the outer region, the profiles collapse very well. Most notably, the turbulent intensity at $z^+ \geq 1000$ shows a very low turbulent intensity, close to zero. **Figure 11** shows the turbulent intensities profiles wall-normal z scaled with boundary layer thickness δ. The boundary layer collapses very well in the outer region, which indicates the boundary layer thickness δ for all profiles is similar at similar U_∞. In summary, the analysis of the variability of the turbulent boundary layer in the spanwise direction demonstrates a high level of agreement. The profiles show acceptable collapse behavior, and the flow quality is very high across the smooth surface.

4. Flow validation and smooth-wall case

An experiment of a wall-normal was performed over the smooth wall at three different free-stream velocities $U_\infty = 5$, 8 and 10 m/s at the exact free-stream location $x = 1.5$ m, and then compared to high-quality experimental data taken from Harun [7]. The taken data are subjected to FPG with different pressure gradient parameters $\beta = -0.52, -0.43, -0.42$, and -0.33. This section analyzes and discusses the diagnostic plot, the inner- and outer-scale of the mean velocity profile and turbulent intensity, the velocity defect for both data and the spectral analysis of the current experiment. **Table 1** shows the experimental parameters for the current study. The

Exp Code	Surface Type	Location $x/l\%$	U_∞ (m/s)	δ (mm)	U_τ (m/s)	$Re\tau$
S	Smooth	50	5.22	87.3	0.2241	1204
S	Smooth	50	8.01	115.2	0.3380	2395
S	Smooth	50	10.27	127.8	0.4250	3410

Table 1.
Experimental parameters for smooth surfaces, $l^+ = lU_\tau/\nu \approx 15 - 30$ for all.

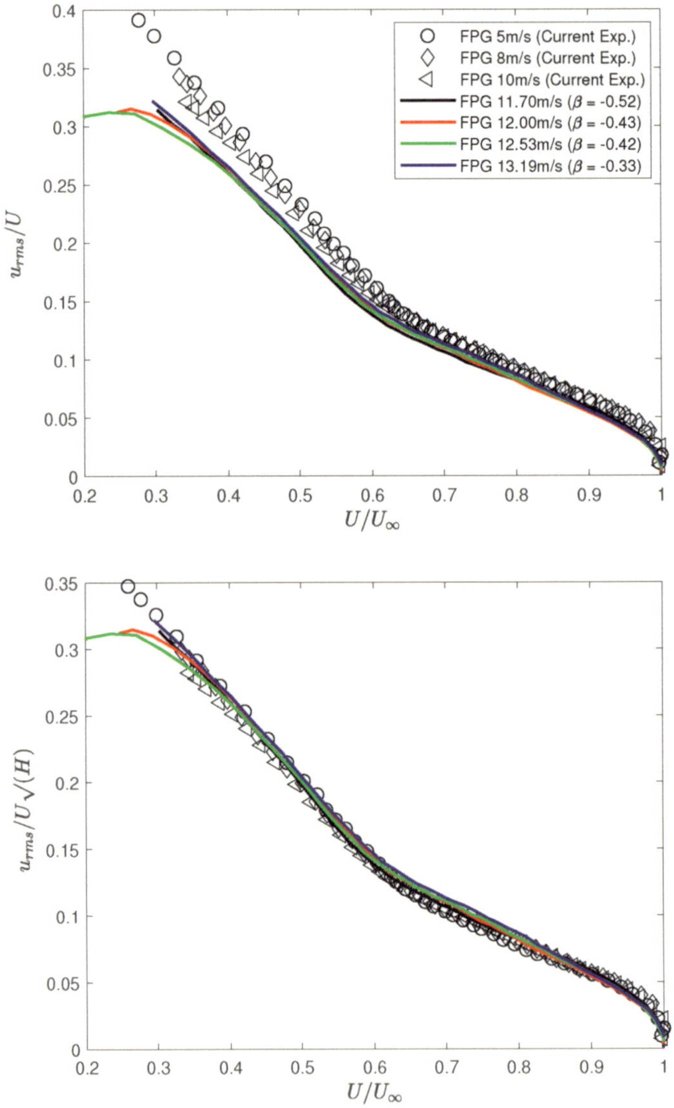

Figure 12.
Comparison of diagnostic plot of the smooth data of the current experiments and FPG data taken from Harun [7], the data presented by the solid colored lines. The top plot shows the classical scaling, and the bottom plot shows the new scaling of diagnostic plot for the pressure gradient data.

current section is divided into three main sections: diagnostic plot, turbulent statistic, and spectral analysis.

4.1 Diagnostic plot

Figure 12 compares the diagnostic plot of the smooth surface and the smooth pressure gradient data taken from Harun [7]. The top figure shows the classical scaling of the diagnostic plot, where the turbulent intensity u_{rms} is scaled with U as a function of U/U_∞. In the top figure, the profile does not collapse well for the entire region due to pressure gradient effects. Recently, Drozdz et al. [25] have introduced a new scaling of the diagnostic plot for the turbulent boundary layer exposed to the pressure gradient effects, where the shape factor H involved in the scaling of u_{rms}/U to be $u_{rms}/\left(U\sqrt{H}\right)$. The new procedure provided a correct pattern for the diagnostic plot for pressure gradient flow, where all profiles collapse very well at $0.8 \leq U/U\infty \leq 0.9$. This is agreed well with the finding of Drozdz et al. [25].

4.2 Turbulence statistics: velocities and turbulence intensity profiles

Figure 13 shows the velocity and turbulent intensity profiles. The current comparison between all profiles of the current experiments is represented by the symbols (∘), (⋄), and (∇) and the colored solid lines represent the data taken from Harun [7]. The current plots are scaled with the skin-friction velocity U_τ as a function of scaled wall position $z^+ = zU_\tau/v$. The velocity profile shows that both data are collapsed very well in the inner region, logarithmic layer in the range of $100 \leq z^+ \leq 0.15Re_\tau$. However, it is not the wake region, where the data have a different Reynolds number Re_τ and is performed in different facilities with different flow environments. The turbulent intensity profiles show the maximum turbulent production at $z^+ = 15$, which follows the standard features of the turbulent boundary layer [6, 7]. However, the data that are taken from Harun [7] show a lower turbulent production, even though the data have a higher Reynolds number Re_τ as it is not supposed to be. When the Reynolds number increases, the near-wall peak also increases [6, 7, 15, 26]. However, the difference in both cases is due to the effect of different wind tunnel environments and spatial attenuation issue. In other words, the hot-wire sensor length l^+ [27].

Furthermore, **Figure 14** shows the mean velocity profile (top) and turbulent intensity profiles (bottom) are scaled with the outer parameter. Where the velocity and turbulent intensity are normalized with free-stream velocity U_∞ and the wall-normal scaled with the boundary layer thickness δ. The result of velocity profiles shows that all profiles collapse well at the wake region $z/\delta = 0.8$. The turbulent intensity results collapse at $z/\delta = 0.8$. This difference is due to different boundary layer thicknesses between both data and the flow test facility.

Figure 15 compares the velocity defect between both experimental data. All the profiles show an excellent collapse for the entire boundary layer regions. All the profiles shifted downward due to the effect of the pressure gradient on the turbulent boundary layer [24, 28]. Overall, all the analyses provided for validating the smooth-wall case over the wind tunnel floor reveal that the turbulent boundary layer has high-quality data and follows the standard features for the

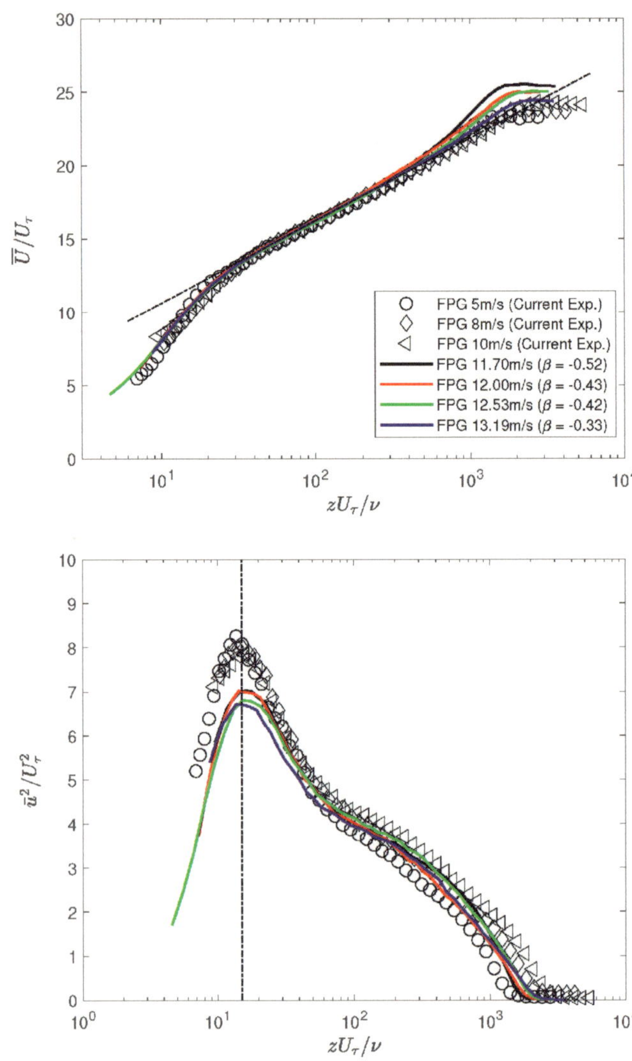

Figure 13.
Comparison of mean velocity and turbulent intestines of the smooth data of the current experiments and FPG data taken from Harun [7], the taken data presented by the solid colored lines. The top plot shows the mean velocity profile, the black dashed-dot line represents the log-low. The bottom plot shows the turbulent intensity, where the black dashed-dot line shows the location on the inner peak at $z^+ = 15$.

boundary layer. All the differences between the current experiments and the referenced data taken from Harun [7] are within the acceptable limit.

4.3 Spectral analysis

A spectral analysis was conducted to evaluate and understand the distribution and significance of various scale features in the turbulent boundary layer in both the near-wall and outer regions. One way to illustrate this is by analyzing all possible length scales that can exist at a particular location with respect to the wall. A pre-multiplied streamwise energy spectra provide a quick way to understand this, as the length scale can be non-dismensionalized with the boundary layer thickness (δ). To start with, it is

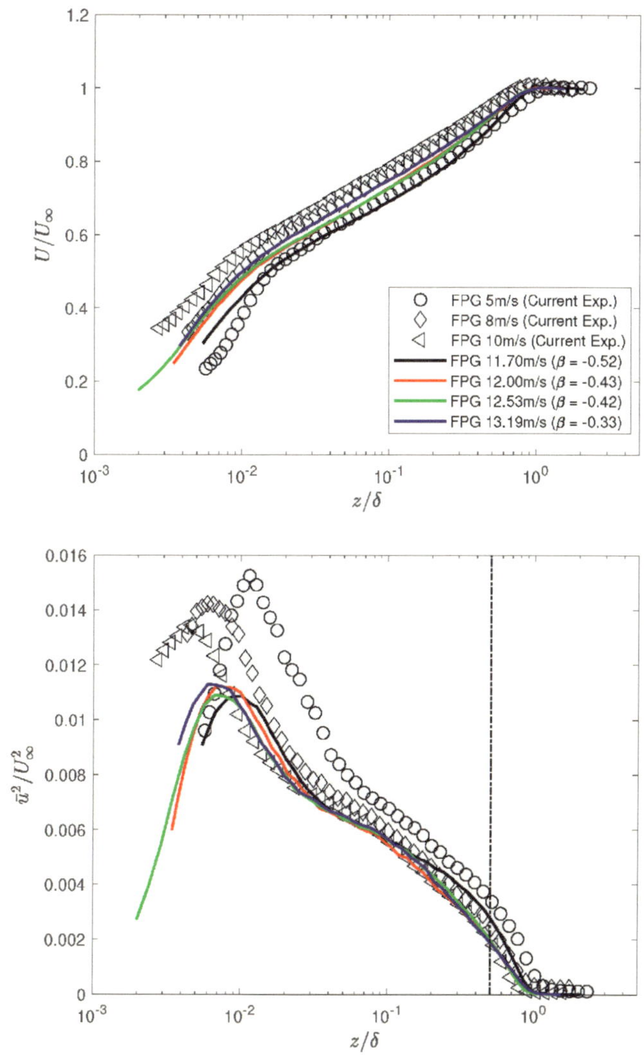

Figure 14.
Comparison of mean velocity and turbulent intestines of the smooth data of the current experiments and FPG data taken from Harun [7], the taken data presented by the solid colored lines. The top plot shows the mean velocity profile and the bottom plot shows the turbulent intensity, where the dashed-dot line represents the outer peak at $z/\delta = 0.5$.

necessary to compute the pre-multiplied streamwise energy spectra $k_x\phi_{uu}$, where the expression k_x represents the streamwise wave number, while ϕ_{uu} represents the spectral density of the streamwise velocity fluctuations. ϕ_{uu} is obtained by a correlation of a velocity fluctuation to itself and this is easily computed in many software using a fast Fourier transform (FFT). Here, $k_x = 2\pi f/U$, where f is the frequency, and U is the local streamwise mean velocity. The velocity spectra are pre-multiplied so that the area under the curve of $k_x\phi_{uu}$ on the semi-logarithmic plot is $\overline{u}^2$.

The streamwise length scale is presented by λ_x, where $\lambda_x = 2\pi/k_x$. Here, the data are normalized with the local friction velocity U_τ, and the wavelength is normalized with $v/U\tau$ and boundary layer thickness δ, as shown in the bottom and top axis. **Figure 16(a)** shows the energy spectra at $z^+ = 15$ in the near-wall and at $z = 0.5\delta$ in the outer region. **Figure 16(a)** shows that the flow at different free-stream velocity $U_\infty = 5$, 8 and 10 m/

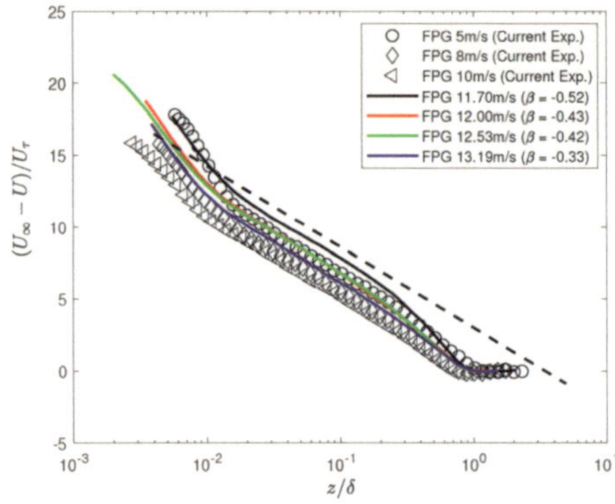

Figure 15.
Comparison of velocity defect profile of the smooth data of the current experiments and FPG data taken from Ref. [7], the taken data presented by the solid colored lines. The black dashed line represent the velocity defect low.

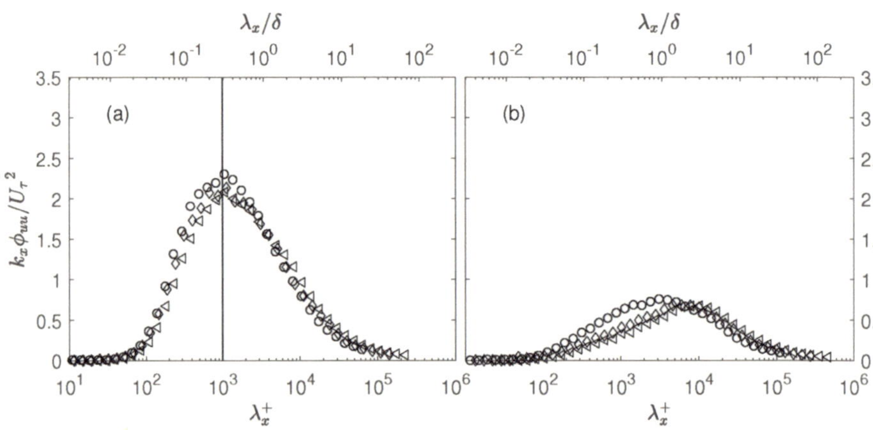

Figure 16.
Individual energy spectral at selected location. (a) $z^+ = 15$ in the near-wall region, and (b) $z = 0.5\delta$ in the outer region. ∘, ⋄, ◁ represents the smooth surface at 5, 8, and 10 m/s, respectively. Vertical line in (a) shows $\lambda_x{}^+ = 1000$.

s, having an energized structure of $\lambda_x{}^+ \approx 1000$. The vertical line in **Figure 16(a)** denotes $\lambda_x{}^+ = 1000$, it can be observed that the peak energy occurs at this location, and there is little variation to the energy distribution across the three velocities.

On the other hand, **Figure 16(b)** shows that the highest energy of the flow structure at the outer region of $z = 0.5\delta$ has a length scale of $1 \leq \lambda_x/\delta \leq 3$. The current results indicate that the turbulent boundary layer at PLSWT has features similar to those of a standard turbulent flow in wind tunnels [7, 15, 26]. Further analysis of the spectra map provides an overview of the energy distribution over the entire boundary layer for all flow profiles. **Figure 17(a)-(c)** represent the spectra map of the entire boundary layer for all three profiles. The spectra maps give details of the most energetic region in the boundary layer represented by symbol (+), which shows the inner peak occurs at $z^+ = 15$. The outer peak presented by the symbol ($\times$) shows a different energy level for all profiles due to the effects of different Reynolds numbers Re_τ. As the Re_τ increase, the chance of observing the outer peak in the outer region becomes

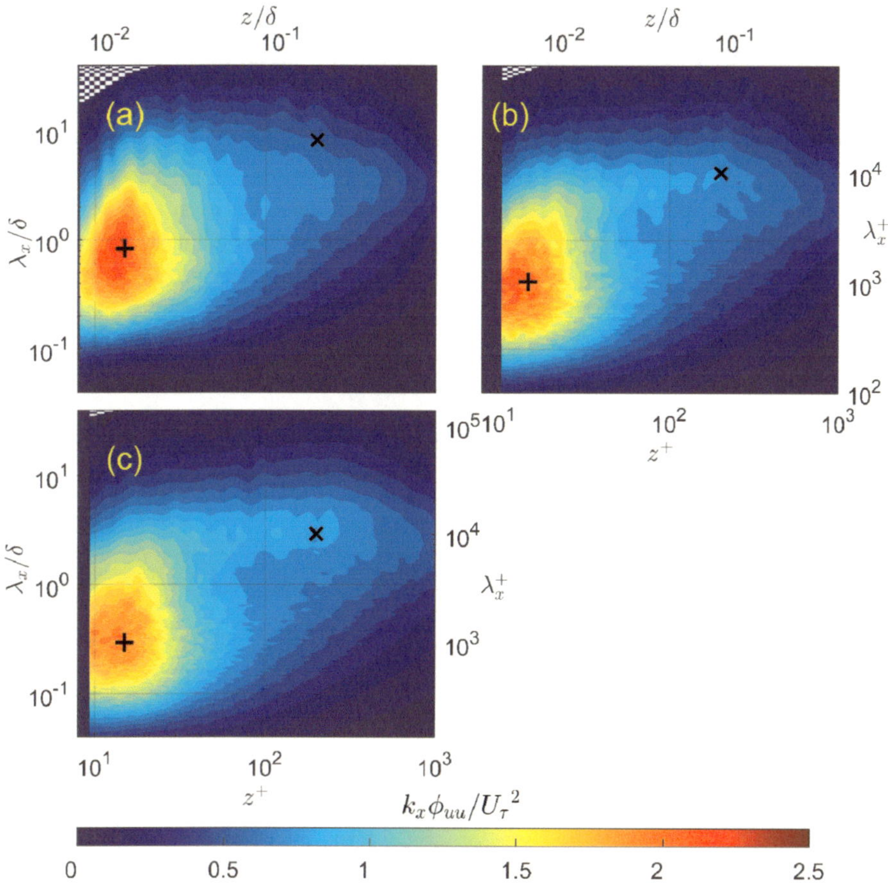

Figure 17.
Spectra maps for the flow over wind tunnel smooth surface (a) 5 m/s, (b) 8 m/s, and (c) 10 m/s. Symbols (+) and (×) represent the near-wall and outer hump locations.

higher as shown in **Figure 17(c)** compared to **Figure 17(a)** and **(b)**, which is due to the effects of large-scale features. Harun [7] shows similar effects when the boundary layer is subjected to a different pressure gradient environment (ZPG, FPG, or APG) or different friction Reynolds number Re_{τ}.

5. Conclusions

Flow quality and turbulent boundary layer experiments have been performed at the Pangkor low-speed wind tunnel (PLSWT). The measurements were carried out using a hot-wire sensor over a flat surface. Different measurement techniques were used to understand and validate the flow in the wind tunnel's test section. The flow uniformity measurement shows excellent results, where the wind tunnel experiences very low-level velocity and turbulent intensity variation due to the effects of the meshing screen and honeycomb device. The turbulence statistical analysis, including diagnostic test, boundary layer velocity profile, outer layer similarity, and turbulent intensity of different spanwise locations, shows a very excellent collapse between all five locations; this indicates that the boundary layer profiles across the spanwise direction do not differ and not suffer from any defect. The analysis of the turbulent boundary layer at the center of the test section at different Re_{τ} shows a good agreement between all three profiles with the compared data. The spectra individual and spectra map analysis shows in detail the turbulent flow features including the streamwise wavelength as well as the location of the most energized region in the boundary layer. The current measurement techniques are useful for researchers working on wind tunnel flow validation. Future measurements of spanwise variability at different streamwise locations and inclination angles are worthwhile.

Acknowledgements

We would like to express our gratitude for the financial support provided by Universiti Kebangsaan Malaysia's Research University Grant GUP-2020-015.

Author details

Zambri Harun*† and Ashraf Amer Abbas†
Department of Mechanical and Manufacturing Engineering, Faculty of Engineering and Built Environment, National University of Malaysia, Bangi, Malaysia

*Address all correspondence to: zambri@ukm.edu.my

† These authors contributed equally.

References

[1] Baals DD. Wind Tunnels of NASA. Vol. 440. Scientific and Technical Information Branch, National Aeronautics and Space ... ; 1981

[2] de Almeida O, Carnevalli F, de Miranda O, Neto F, Saad FG. Low subsonic wind tunnel-design and construction. Journal of Aerospace Technology and Management. 2018;**10**

[3] Britcher C, Landman D. Wind Tunnel Test Techniques: Design and Use at Low and High Speeds with Statistical Engineering Applications. Academic Press; 2023

[4] Scheiman J, Brooks JD. Comparison of experimental and theoretical turbulence reduction from screens, honeycomb, and honeycomb-screen combinations. Journal of Aircraft. 1981; **18**(8):638-643

[5] Harun Z, Wan AW, Ghopa SA, Izhar Ghazali M, Abbas AA, Rasani MR, et al. The development of a multi-purpose wind tunnel. Jurnal Teknologi. 2016;**78**: 6-10

[6] Harun Z, Monty JP, Mathis R, Marusic I. Pressure gradient effects on the large-scale structure of turbulent boundary layers. Journal of Fluid Mechanics. 2013;**715**:477-498

[7] Harun Z. The structure of adverse and favourable pressure gradient turbulent boundary layers [PhD thesis]. 2012

[8] Cattafesta L, Bahr C, Mathew J. Fundamentals of wind-tunnel design. In: Encyclopedia of Aerospace Engineering. Wiley Online Library; 2010. pp. 1-10

[9] Mathew J, Bahr C, Carroll B, Sheplak M, Cattafesta L. Design, fabrication, and characterization of an anechoic wind tunnel facility. In: 11th AIAA/CEAS Aeroacoustics Conference. Monterey, California: AIAA Aerospace Research Central; 2005. p. 3052

[10] Wiriadidjaja S, Hasim F, Mansor S, Asrar W, Rafie ASM, Abdullah EJ. Subsonic wind tunnels in Malaysia: A review. Applied Mechanics and Materials. 2012;**225**:566-571

[11] Harun Z, Abbas AA, Mohammed Dheyaa R, Ghazali MI. Ordered roughness effects on NACA 0026 airfoil. IOP Conference Series: Materials Science and Engineering. 2016;**152**:012005

[12] Hutchins N, Choi K-S. Accurate measurements of local skin friction coefficient using hot-wire anemometry. Progress in Aerospace Sciences. 2002;**38** (4-5):421-446

[13] Monty JP, Harun Z, Marusic I. A parametric study of adverse pressure gradient turbulent boundary layers. International Journal of Heat and Fluid Flow. 2011;**32**(3):575-585

[14] Harun Z, Abbas AA, Lotfy ER, Khashehchi M. Turbulent structure effects due to ordered surface roughness. Alexandria Engineering Journal. 2020; **59**(6):4301-4314

[15] Hutchins N, Marusic I. Evidence of very long meandering features in the logarithmic region of turbulent boundary layers. Journal of Fluid Mechanics. 2007;**579**:1-28

[16] Monty JP, Hutchins N, Ng HCH, Marusic I, Chong MS. A comparison of turbulent pipe, channel and boundary layer flows. Journal of Fluid Mechanics. 2009;**632**:431-442

[17] Henrik Alfredsson P, Örlü R. The diagnostic plot—A litmus test for wall

bounded turbulence data. European Journal of Mechanics-B/Fluids. 2010; **29**(6):403-406

[18] Henrik Alfredsson P, Segalini A, Örlü R. A new scaling for the streamwise turbulence intensity in wall-bounded turbulent flows and what it tells us about the "outer" peak. Physics of Fluids. 2011; **23**(4). Article Number: 041702

[19] Henrik Alfredsson P, Örlü R, Segalini A. A new formulation for the streamwise turbulence intensity distribution in wall-bounded turbulent flows. European Journal of Mechanics-B/Fluids. 2012;**36**:167-175

[20] Henrik Alfredsson P, Segalini A, Örlü R. The diagnostic plot—A tutorial with a ten year perspective. In: iTi Conference on Turbulence. Springer; 2021. pp. 125-135

[21] Clauser FH. Turbulent boundary layers in adverse pressure gradients. Journal of the Aeronautical Sciences. 1954;**21**(2):91-108

[22] Clauser FH. The turbulent boundary layer. Advances in Applied Mechanics. 1956;**4**:1-51

[23] Nagib HM, Chauhan KA. Variations of von Kármán coefficient in canonical flows. Physics of Fluids. 2008;**20**(10). Article Number: 101518

[24] Townsend AAR. The Structure of Turbulent Shear Flow. Cambridge University Press; 1976

[25] Dróżdż A, Elsner W, Drobniak S. Scaling of streamwise Reynolds stress for turbulent boundary layers with pressure gradient. European Journal of Mechanics-B/Fluids. 2015;**49**:137-145

[26] Hutchins N, Marusic I. Large-scale influences in near-wall turbulence. Philosophical Transactions of the Royal Society A: Mathematical, Physical and Engineering Sciences. 2007;**365**(1852): 647-664

[27] Harun Z, Isa MD, Rasani MR, Abdullah S. The effects of spatial resolution in turbulent boundary layers with pressure gradients. Applied Mechanics and Materials. 2012;**225**: 109-117

[28] Townsend AA. The Structure of Turbulent Shear Flow. Cambridge: Cambridge University Press; 1956

Chapter 2

Novel Approach for Turbulent Flow- and Onset Analysis

Mattias K. Gustavsson

Abstract

It is a well-known fact that the matching of experimental data to turbulence models have hitherto not been successful. An example of this is the inability to theoretically predict the Re number at which turbulence onset (transition) occurs. In this paper, some advantages of adopting a "far-from-equilibrium" irreversible process analysis are demonstrated: To illustrate, one may at a single geometric point near a solid wall, compute conditions for mass conservation, 1st, and 2nd laws of thermodynamics – assuming either Newton's viscosity law- or an alternative far-from-equilibrium fundamental model to be valid. While these conditions generally differ for various flows, it is observed that these conditions numerically match each other at Re_D around 2300 for a fully developed pipe flow, or at Re_x between 5×10^5 to 3×10^6 in a developing flat-plate boundary layer flow. This suggests that turbulence onset can be correctly predicted using the novel approach. Criteria and recommendations for experimental flow measurements, i.e. testing conditions, within a proposed far-from-equilibrium zone (e.g. viscous sublayer) is discussed as well.

Keywords: fundamental model, turbulence, onset of turbulence, transition, pipe flow, flat-plate boundary layer, far-from equilibrium

1. Introduction

According to [1], the following question is posed: "Why is turbulence so hard to solve?" It is stated that: "An example of why turbulence is said to be an unsolved problem is that we can't generally predict the speed at which an orderly, non-turbulent ("laminar") flow will make the transition to a turbulent flow." Perhaps, an understanding of the mechanisms resulting in the onset of turbulence is a key issue for the understanding.

The turbulent flow differs from a well-ordered laminar shear flow [2–7], and for a wall-bounded turbulent flow, several near-wall regions appear. Moreover, the kinetic energy dissipation is a rudimentary quantity, associated with the far-from-equilibrium process, and this increases drastically with the onset of turbulence.

A theory that in the following is referred to as the "Classical Turbulence Theory (CTT)," has been developed since the early 20th century. Originally, turbulence was considered a random process [8], involving swirling or rotating fluid elements, commonly referred to as "eddies" or "vorticities." Through the last 50 years, many researchers have focused their interest on the transition and the flow behaviour in the innermost region of a turbulent boundary layer, e.g. coherent structures, bursts, and sweeps [3, 9–14].

The Navier-Stokes relations are originally based on the continuum hypothesis [3] and assuming the flow processes to be governed by Newton's viscosity law (a near-equilibrium irreversible thermodynamic process) [15]. This set of relations can be excellently applied on laminar flows, however, this set of relations is also assumed to be valid in turbulent flows [15–21]. Commonly, in CTT, a Reynolds decomposition is made, allowing for a simplification of the Navier-Stokes relations. The ensuing Reynolds-averaged Navier-Stokes relations incorporate new (unknown) quantities referred to as "Reynolds Stresses." This is the so-called "closure problem" of turbulence which requires a modelling of the stresses to account for temporal- and spatial scale influence.

In CTT, a constant shear stress across the inner part of the turbulent boundary layer is assumed. This represents an empirical assumption: In short, it is assumed (cf. Hinze [22] p. 468) that both viscous and turbulent interactions contribute to the shear stress, a sum which is constant across all these regions, cf. [5] (and discussed partly in Section 2.3). According to Panton [3] (p. 720): "At the wall itself, the stress τ_0 is 100% viscous shear because the fluctuation must vanish. Through the buffer layer τ decreases while the Reynolds stress $-\rho\overline{u_1 u_2}$ increases. At the outer edge of the buffer layer, the stress is essentially all turbulent Reynolds stress and is equal to the wall value τ_0. The Reynolds stress continues to be constant through the overlap region." According to [5], the viscous processes have a negligible influence in the log-law region. The value assigned to the Reynolds Stress in the log law region is $-\overline{u_1 u_2} = (U^*)^2$, cf. Section 2.3.

A novel approach to analyse turbulent flows is presented here, see [23, 24], and it is based on the onset of a far-from-equilibrium [25–27] fundamental process at a specific threshold condition. Hence, a similar large-scale flow behaviour is obtained as in the CTT approach. However, some differences may be noted:

- The novel approach puts focus on the fundamental model as responsible for the build-up of the near-wall zones; viscous sublayer, buffer layer, overlap region/log-law region, and outer region of a turbulent wall boundary layer, which occurs due to a kinetic energy dissipation near the solid wall.

- On the nature of here so-called "visible swirls" or "visible eddies" (as all fluid elements within the novel approach falls under the category of "eddies"), the novel approach suggests conditions where these may form – however, the influence of these swirls and eddies appear to be limited when computing the net kinetic energy dissipation.

In this paper an attempt is made to discuss the application of the novel approach. Section 2 presents a simplified discussion on how a far from equilibrium process differs from the traditional continuum approach. Then, a thorough description of the novel fundamental far-from-equilibrium process is given and applied in computations of the shear stress variation in a fully developed pipe flow.

In Section 3 the discussion is extended to the analysis of turbulence onset (transition) in pipe flows. Noteworthy is that for the novel approach, it is possible to extract the onset condition by considering the mass conservation and 1st and 2nd laws of thermodynamics at a specific Re_x position along a flat wall, or a specific Re_D flow condition for a pipe flow. Both these Re numbers agrees well with established experimental data.

Section 4 is devoted to a discussion on the planning of near-wall experiments and simulations, incorporating far-from-equilibrium considerations.

2. Novel approach

2.1 Comparisons between irreversible continuum- and far-from-equilibrium processes: A summary of differences

A difference between continuum CTT approach and the far-from-equilibrium process, grossly simplified, is the following:

- For the continuum processes the principle of local equilibrium is applicable, that is at all positions (y_1, y_2, y_3, t), quantities like temperature, pressure, density all have a unique value. These quantities are defined locally in an elemental volume with a large enough number of molecules allowing for the principles of thermodynamics to be valid, i.e. the viscous dissipation of a soft gradient shear-flowing Newtonian fluid can in an elemental volume be strictly computed by knowledge of the local flow gradient.

- In the far-from-equilibrium process, there is an overall mechanism, which controls the global behaviour. Often ordered structures (e.g. waves) are created, referred to as "dissipative structure," and results in a significantly larger net entropy generation. One such mechanism is the MEP (Maximum Entropy Production principle) [28], which organises local flow behaviour. The build-up of viscous sublayer, buffer layer and log-law region are results of this organisation, see [23]. An important aspect of this condition is that – in contrast to continuum processes – a specific gradient in the flow does not necessarily translate into a kinetic energy dissipation [23–24]. To illustrate, a flow gradient within the viscous sublayer, which is considered a saturated kinetic energy dissipation zone, the far-from-equilibrium model will give a strict number on the kinetic energy dissipation per unit volume. However, in zones where "defects" are located, or "clouds of defects," the local condition with flow gradient can have any local kinetic energy dissipation, from 0 up to, but not exceeding, the kinetic energy dissipation for a corresponding saturated condition. This aspect of a far-from-equilibrium process is important, as it allows the overall invisible MEP process to organise the dissipative structures, while less dissipative zones will immediately adapt to the local conditions (without any violation of the principle of mass conservation). At low-dissipative zones, the flow may switch back to continuum mode. (Nonetheless, the return to continuum may incorporate some hysteresis phenomena, as discussed in [24].)

2.2 Method

The proposed far-from-equilibrium model valid after turbulence onset, is in [23] expressed in terms of a kinetic energy dissipation $\frac{\mathrm{d(ke)_{res}}}{\mathrm{d}t} = C_A \frac{\rho L}{\delta} U_{\mathrm{slip}}$, which is an Eulerian expression.

The background stems comparing the original concept of a solid with that of a fluid: A solid beam obeys the rules of shear stress *vs.* strain, principally like a laminar fluid flow where instead of strain the strain rate is introduced.

Consider a glass solid in the shape of a beam, but where a far-from-equilibrium impact of, say, a bullet, results in a bullet hole in the beam together with a creation of a spider-network-type of fractures around the hole. Some immediate observations on the far-from-equilibrium impact process can be observed: First, the concentration of fractures decrease with increasing distance from the bullet hole, and second, the appearance of the spider-web-type of fractures is not at all random – an ordered dissipative structure may have been active in the creation of the web of fractures around the bullet hole.

Now, if applying the same bending operation of the bullet-fractured beam, with the aim of recording the overall shear stress vs. strain, the results will be completely different. A solid mechanics scientist would state that this analysis is invalid, and if attempting to understand the solid mechanics behaviour of a fractured beam, it would be of more interest to study the resulting slippage along a single fracture and connect the slippage with some kind of friction.

If converting this drastic idea of fractures into a Newtonian fluid, it is immediately realised that the corresponding concept to a solid fracture is that of a slip flow. In [23], it was argued that if this assumption is made on slip flow occurring in case a far-from-equilibrium irreversible process onsets, then with some basic mathematically and physically realistic assumptions on the connection between a local slip length and a local slip velocity to be directly proportional, i.e. $L = C_B U_{\text{slip}}$ (**Figure 1**).

Instead of fractures originating from a bullet hole radially in a solid, consider the assumption for a turbulent wall-bounded flow, where instead square flakes originating from the wall, with slippage U_{slip} in the y_1-direction, where the flake thickness is δ, flake length is L and flake width is K. One can then organise the L and δ to reproduce the experimental viscous sublayer, buffer layer, and log-law regions (**Figure 1**).

The friction force of a single flake can be replaced by a shear stress acting on the flake, $\tau = F_1/A = F_1/(K \cdot L)$, which gives a kinetic energy dissipation rate per unit volume $\mathrm{d(ke)}_{\text{res}}/\mathrm{d}t = \tau \cdot U_{\text{slip}}/\delta$ locally within the flow when ensemble averaging.

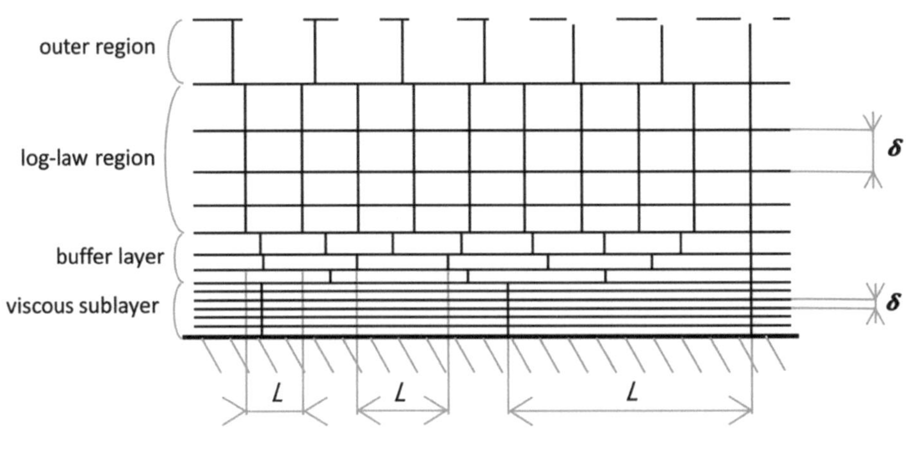

Figure 1.
Fracture structure for a turbulent fluid flow parallel to a solid wall, resulting in fully developed viscous sublayer, buffer layer and log-law zones. Note that organisation of L and δ totally depend on the experimentally recorded velocity vs. y_2^+ *(except resolution). Locally, the flow has the same kinetic energy dissipation and same velocity gradient as long as the ratio of L vs. δ is the same, cf. [23].*

A second assumption is that $\tau \propto U_{\text{slip}}$. From this, one may postulate: $\tau = C_A \cdot \rho \cdot C_B \cdot U_{\text{slip}}$, which gives:

$$\frac{\mathrm{d(ke)}_{\text{res}}}{\mathrm{d}t} = C_A \frac{\rho L}{\delta} U_{\text{slip}} \tag{1}$$

Similarly to the case with the fractures in the bullet hole, the local kinetic energy dissipation is the highest in the viscous sublayer, where the slip length L is the longest, together with the fracture thickness δ is the smallest. Immediately, with the present formulation, it is found that for fully developed turbulent pipe flows at $\mathrm{Re}_D \geq 5000$, the net kinetic energy dissipation in the viscous sublayer is larger than the remaining net kinetic energy dissipation throughout all the other zones together. In addition, within the viscous sublayer zone, there is no eddy activity, at all.

The action of the far-from-equilibrium process along a viscous sublayer, can be illustrated with the example of an erosion rippling process [29], caused by an impacting jet stream of round hard particles impacting a ductile solid surface at a low angle of attack – described in detail in [23]: It is argued that a far-from-equilibrium slip-roll zone is built up closest to the wall. Two other zones can be identified, one zone with no far-from-equilibrium process occurring at all, which represents the major part of the entire flow volume. An intermediate zone also exists, which is a mixture of both the local zones of far-from-equilibrium processes being onset, as well as local offset zones. It has been identified that prior to the rippling process initiating, the reflecting flow is "laminar" [29], while when the rippling process has onset, the reflecting flow is "turbulent" [29]. Hence, for this case, the observation of a turbulent behaviour is a clear indication, or symptom, on the initiation of an erosion rippling process at the ductile wall surface [23].

2.3 Variation of shear stress across a turbulent boundary layer

By organising the time-averaged mean velocity variation in accordance with experiments as proposed in [23], and applying a novel fundamental model, a well-organised kinetic energy dissipation rate across the boundary layer appears to occur, cf. Figure 7 in [23]. Hence, the viscous sub-layer, buffer layer, and the log-law regions may seem to occur because of a self-organised grouping of fracture cells (to maximise the overall entropy production of the turbulent flow), cf. Section 2.6 in [23].

In this grouping, it becomes evident that the viscous sublayer represents a region of saturated – i.e. maximum entropy generation per unit volume, or equivalently, a maximum kinetic energy dissipation per unit volume. The significant amount of net kinetic energy dissipation within the viscous sublayer region, which may account for more than 50% of the total net kinetic energy dissipation of the total flow, immediately suggests that the specific thickness of the sublayer must have significant effect on the net kinetic energy dissipation, and that this thickness variation of the sublayer needs to be incorporated in the analysis [23].

Section 2.7 in [23] describes an approach in which the fracture model may account for the effects of various surface roughnesses, which is consistent with experimental observations on increasing of the U_1^+ intersection point of the log-law relation in the y_2^+ axis. (The Moody diagram also indicates an increased pressure drop [24] for situations when the surface roughness is higher.) A model is proposed, on how the C_A and L parameters vary with position from the wall, for different wall surface roughness conditions, cf. Section 2.7 in [23], and Figures 8–9 in [23]. This model can be simplified into a model describing the conditions in the viscous sublayer $C_A L_{\text{max}}$, by which all other conditions throughout the turbulent wall boundary layer can be obtained in a non-dimensional manner, cf. Eqs. 17 and 18 in [23].

Also, with this novel fundamental model, it becomes immediately obvious why the generation of different zones – viscous sublayer, buffer region, log-law (or overlap) region, as well as the outer region of a turbulent boundary layer develops:

- Viscous sublayer: saturated region, constant $L = L_{\text{max}}$ and constant δ.

- Buffer region, varying L and varying δ. Transitional region between log-law and viscous sublayer, not saturated, hence defects may occur, resulting in visible swirls (visible/recordable eddies). Due to varying L and δ as compared to neighbouring cells, a greater likelihood of creation of visible swirls (buffer region is identified as a "turbulence" (or visible eddy) production region).

- Log-law region: constant $L \approx L_{\text{max}}/6$, but varying δ. Not saturated, hence defects may occur, resulting in visible swirls (visible/recordable eddies)

- Outer region: a region with only occasional presence of onset novel fundamental zones. In the interface between novel fundamental zone and Newtons viscosity law zone, results in large swirly behaviour – which is understood as "large turbulent eddy" behaviour or "large-scale turbulence" by the external observer.

This organisation of fracture cells across a turbulent wall boundary layer, results in a corresponding principal shear stress variation (not to scale) for a fully developed turbulent pipe flow which can be depicted in **Figure 2**.

The corresponding principal shear stress variation (not to scale) for the CTT case at fully developed conditions is depicted in **Figure 3**, while shear stress for a laminar-flow case at fully developed conditions is depicted in **Figure 4**.

The shear stress behaviour as computed by the novel approach across the viscous sublayer is constant but differ in magnitude as compared to CTT. This represents the main difference between CTT and the novel approach: The shear stress according to

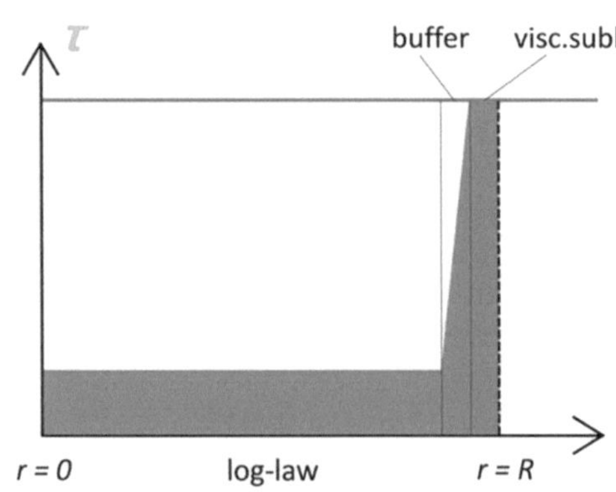

Figure 2.
Shear stress across a pipe cross-section according to novel approach, where $L = L_{\text{max}}$ within the viscous sublayer is approximately 6 times larger than L in the log-law region. Shear stress across all these regions is governed by the relation $\tau = \rho C_A L$. The action of the novel approach fundamental mechanism is coloured in green.

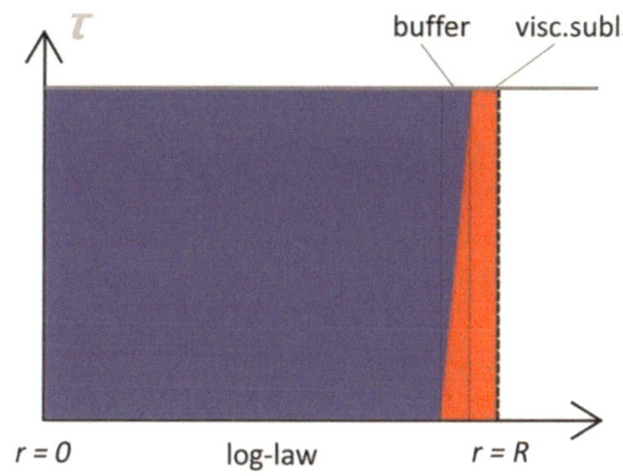

Figure 3.
Shear stress across a pipe cross-section according to CTT, where viscous processes (their action marked in red colour) account for 100% of shear stress within the viscous sublayer, and Reynolds stresses (their action marked in blue colour) account for 100% of shear stress within the log-law region. The buffer layer shear stress is a combination of these two shear stress mechanisms.

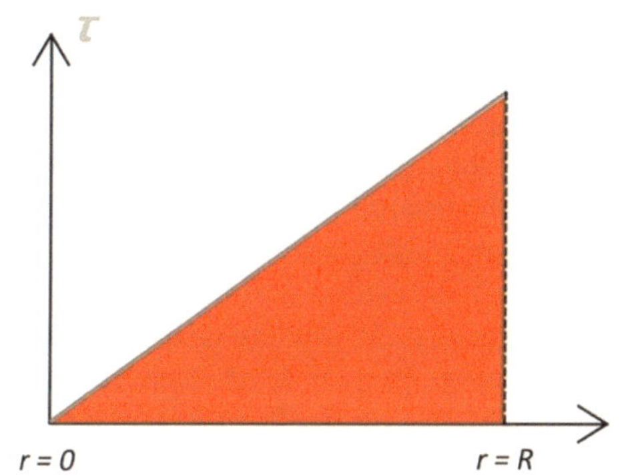

Figure 4.
Shear stress across a pipe cross-section for laminar flow (at fully developed laminar pipe-flow conditions), where the shear stress across all these regions is governed by the relation $\tau = -\mu \frac{\partial U_1}{\partial r}$. The fundamental mechanism based on Newton's viscosity law describes the action of the viscous processes (their action marked in red colour) which account for 100% of shear stress within the entire pipe cross section.

the novel approach is approximately 6 times higher than the shear stress computed for the log-law region.

From a modelling point of view, the parameter $C_A L$ takes the role of the Reynolds stress $-\overline{u_1 u_2}$ within the log-law region, which according to CTT is constant throughout the low-law region. Note that $C_A L$ is according to the novel approach also constant throughout the log-law region, cf. Section 2.6 in [23]. The shear stress expressions (for pipe flows) in the various zones are summarized in **Table 1**.

3. Flow modelling

3.1 Test case - fully-developed turbulent pipe flows

A computation of 20 different turbulent flow cases of fully developed pipe flows, for air flow at 20 °C in a 14 cm diameter tube, with different surface roughnesses, arrives at results according to **Table 2**.

Zone	Novel approach	CTT	Laminar flow
viscous sublayer	$\tau = \rho C_A L$ $(= \rho C_A L_{max})$	$\tau = -\mu \frac{\partial U_1}{\partial r}$	$\tau = -\mu \frac{\partial U_1}{\partial r}$
buffer layer	$\tau = \rho C_A L$	$\tau =$ part$(-\rho \overline{u_1 u_2})$ + part$(-\mu \frac{\partial U_1}{\partial r})$	$\tau = -\mu \frac{\partial U_1}{\partial r}$
log-law region	$\tau = \rho C_A L$	$\tau = -\rho \overline{u_1 u_2}$	$\tau = -\mu \frac{\partial U_1}{\partial r}$
outer region	$\tau =$ part$(\rho C_A L)$ + part$(-\mu \frac{\partial U_1}{\partial r})$	$\tau \to 0$	$\tau = -\mu \frac{\partial U_1}{\partial r}$

Table 1.
Expressions for the variation of the shear stress in a laminar and a turbulent pipe flow. CTT assumes a constant shear stress, while the novel approach relies on a time-averaged mean velocity profile.

According to **Table 2**, the viscous sublayer and buffer regions contain most of the kinetic energy dissipation occurring. Most importantly, results with this novel approach arrives at a finding that the zones with turbulence generation and large eddies have a low, marginal, influence on the net kinetic energy dissipation.

In Section 3.4 in [24] a discussion is offered, on estimating the influence from visible eddies themselves, arriving at numbers as low as 1–3% influence caused by the visible eddies as compared to the overall flow. In present novel approach, it is zones with locally reduced kinetic energy dissipation which cannot maintain the onset of the novel fundamental slip flow, which results in imbalances of forces, which in turn may result in the onset of a swirling eddy. As discussed in Section 2.1, the presence of defect zones is a natural occurrence in a far-from-equilibrium large-scale ordered structuring, which is why swirls may occur. In contrast, also as stated in Section 2.1, any such "defect" zones cannot exist in a near-thermodynamic equilibrium irreversible continuum process governed by Newton's viscosity law, hence swirls cannot be created in this way.

In Section 2.9 in [23], the analysis of turbulence onset is offered. A recent analysis of the onset/offset between laminar flows and the slip-fracture structure for fully developed pipe flows, arrived at onset/offset Re_D-numbers in fair agreement with corresponding experiments, i.e. around 2300 for the situation the wall surface roughness is smooth ($\varepsilon/D = 0.00$) and if assuming a slightly (50%) higher velocity gradient in the onset position (in the vicinity of the solid wall) in the laminar-flow side of the turbulent spot [3, 30–33].

3.2 Flat plate developing boundary-layer flow

3.2.1 Traditional description of boundary layer flow along a flat plate

The literature describes a boundary layer flow along a flat plate, which initially is laminar, and at a critical position x_{cr}, onsets a turbulent flow boundary layer. For both the laminar zone of the boundary layer, and the turbulent zone of the boundary layer, it has been experimentally found that the thickness of the boundary layer increases with downstream position x.

The literature presents various ways of analysing these flows. For the laminar flow zone without a pressure gradient along a plate, the Blasius relations (cf. Section 7.4 in [2])

Re	f	τ_{wall} (N m^{-2})	ε/D	U_{mean} (m s^{-1})	$C_A L_{max}$ (m^2 s^{-2})	% of $\frac{d(KE)}{dt}$ in v. subl.	% of $\frac{d(KE)}{dt}$ in buffer r.	% of $\frac{d(KE)}{dt}$ in log-law
5000	0.03739	0.001638	0	0.5393	0.002103	60.0	36.1	3.9
5000	0.03745	0.001641	0.00005	0.5393	0.002105	60.0	36.1	3.9
5000	0.03795	0.001663	0.0005	0.5393	0.002114	59.8	36.2	3.9
5000	0.04261	0.001867	0.005	0.5393	0.002178	58.2	37.9	3.9
5000	0.07595	0.003327	0.05	0.5393	0.002695	82.8	17.2	0.0
1×10^4	0.03088	0.005415	0	1.079	0.007380	58.2	35.9	5.9
1×10^4	0.03096	0.005429	0.00005	1.079	0.007386	58.2	35.9	5.9
1×10^4	0.03164	0.005549	0.0005	1.079	0.007445	58.0	36.0	6.0
1×10^4	0.03763	0.006599	0.005	1.079	0.008109	58.0	38.8	3.2
1×10^4	0.07380	0.01294	0.05	1.079	0.01052	82.6	17.4	0.0
1×10^5	0.01799	0.3155	0	10.79	0.5251	54.6	34.5	8.5
1×10^5	0.01826	0.3202	0.00005	10.79	0.5290	54.6	34.5	8.5
1×10^5	0.02033	0.3565	0.0005	10.79	0.5587	54.7	34.8	8.4
1×10^5	0.03131	0.5491	0.005	10.79	0.7179	56.6	38.6	4.8
1×10^5	0.07178	1.259	0.05	10.79	1.028	82.4	17.6	0.0
1×10^6	0.01165	20.43	0	107.9	41.39	53.6	33.9	8.7
1×10^6	0.01265	22.18	0.00005	107.9	43.14	53.6	34.0	8.7
1×10^6	0.01721	30.18	0.0005	107.9	50.66	54.0	34.4	8.7
1×10^6	0.03047	53.43	0.005	107.9	71.14	56.9	38.9	4.2
1×10^6	0.07157	125.5	0.05	107.9	102.5	82.4	17.6	0.0

Table 2.
Computations. The net integrated kinetic energy dissipation across a 1 m pipe, across the radius, using Eq. 18 in [23], arrives at a number multiplied by $C_A L_{max}$. This number is compared to the corresponding experiments computed using eq. 13 in [23]. The $C_A L_{max}$ coefficient is determined resulting in the net kinetic energy dissipation of the computations matching the corresponding experiments – i.e. a 1st law balance agreement between experiments and computations with novel approach model. Time-averaged mass flow rates in full agreement. Time-averaged velocity profile is in full agreement. (originally presented in [24]).

allow the computation of the boundary layer thickness and wall shear stress – which varies along the downstream position x.

3.2.2 Applying the proposed theory of turbulence onset for a developing boundary-layer flow

Regarding the turbulence onset, and possibilities to estimate this, consider the original set of equations (Eqs. 20–21 in [23]), which described the local conditions for onset in a stationary pipe flow by equalling both the local kinetic energy dissipation for the two fundamental models, as well as simultaneously equalling the velocity gradient. Combining these two equations, and considering the velocity gradient in the vicinity of the solid wall, the condition for onset can be expressed as:

$$C_A L_{max}\Big|_{onset} = \nu \frac{\partial U_1}{\partial y_2}\Big|_{onset} \tag{2}$$

Noteworthy is that this relation *only states the condition at the onset position* along a solid wall.

Onset of turbulence in a fully developed pipe flow at stationary flow conditions:

For turbulence onset in a stationary pipe flow, the behaviour at steady-state stationary flows is recorded, across a certain section (which can be comparatively short), assuming the flow has fully developed. Hence, the possible distance from flow inlet to the section of testing, i.e., the distance required to reach fully developed flow conditions, could be hundreds of diameters in length, or more. The net flow rate is then adjusted, whereby a new steady-stationary flow is developed for this new flow rate condition. However, around a critical flow velocity, between the laminar flows and turbulent flows, if controlling the flow carefully, it is sometimes possible to arrange a flow allowing the observation of stationary turbulent spots building up along the wall at fixed positions. Also, one may ascertain their spreading around the circumference at a fixed downstream position, after which the turbulent flow is onset.

Onset of turbulence along a non-developed boundary layer:

Consider a steady-state observation of **Figure 5**, which represents the relevant observation from the z (or y_3) direction. Although the flow observed from the z direction appears stationary, the flow should not be considered fully developed at any position along the plate, i.e., along the x-direction (y_1-direction), as the boundary-layer behaviour varies along the plate.

In the following example, it is assumed that the boundary layer along a flat plate develops also in a similar manner in a pipe flow with initial laminar inflow at the beginning of the pipe. Hence, an attempt is made to connect the onset for several different pipe flows, using Eq. 2.

3.2.3 Estimating $C_A L_{max}$ *values at the turbulence onset condition from empirical observations*

To verify the theoretical estimation Eq. 2 from the fracture-structure approach, an attempt is made to estimate the range of $C_A L_{max}$ values corresponding to the Re_x numbers between 5×10^5 and 3×10^6, i.e., the empirically found range for turbulence onset.

From an empirical finding [24][1], it is fair to assume the following at the onset position:

$$C_A L_{max} \propto (U_{max})^2 \text{ or } C_A L_{max} = \xi (U_{max})^2 \tag{3}$$

At the onset position, assume that the velocity gradient in the vicinity of the wall is approximately:

$$\left.\frac{\partial U_1}{\partial y_2}\right|_{\text{near wall}} \approx 10 \frac{U_{max}}{\Delta} \tag{4}$$

where Δ is the local thickness of the boundary layer (at the onset position).

[1] In [24], a connection was established as $C_A L_{max}$ proportional to ${U_{mean}}^{1.87}$ at fully developed pipe flow conditions, when $\varepsilon/D = 0$. (The power of 1.87 increased with increasing surface roughness, up to 1.99 for $\varepsilon/D = 0.05$, for reasons discussed in Section 3.5 in [24]).

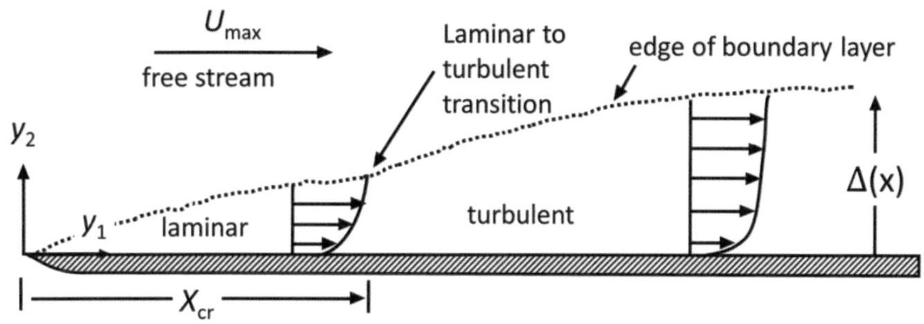

Figure 5.
Traditional description of a boundary layer flow developing at a flat plate, with onset of turbulence occurring at a critical position x_{cr}. The thickness Δ of the boundary layer in turbulent state develops more rapidly than the laminar flow boundary layer thickness, with increasing downstream position x.

Introducing Eqs. 3 and 4 into Eq. 2 then gives the following relation at the onset position:

$$\frac{10x}{\Delta} \approx \xi \, \mathrm{Re}_x \tag{5}$$

The boundary-layer thickness at the onset position can be estimated from the laminar-flow boundary layer thickness (Blasius formula), cf. [2]:

$$\Delta = \frac{5x}{\sqrt{\mathrm{Re}_x}} \tag{6}$$

Combining Eqs. 5 and 6 gives at the onset position:

$$\xi = \frac{2}{\sqrt{\mathrm{Re}_x}} \tag{7}$$

which for onsets at Re_x numbers between 5×10^5 and 3×10^6 corresponds to onsets at ξ numbers between 0.0011547 and 0.0028284. In turn, the corresponding range of $C_A L_{\max}|_{\text{onset}}$ for various pipe flows can be estimated using Eq. 3.

First iteration:

Is it possible to directly estimate the $C_A L_{\max}|_{\text{onset}}$ at the onset position, and to compare this with the range of $C_A L_{\max}|_{\text{onset}}$ computed from the range of Re_x numbers at which turbulence is considered to onset?

Without information on variations along the boundary layer, one important parameter that may be useful is the concept of friction velocity – originally defined as a scaling velocity for a turbulent boundary layer. This concept can be extended in the following, as it is in the traditional turbulence sciences stated to not represent any real velocity. However, for the fracture structure model, the friction velocity can be demonstrated to equal the slip velocity in the viscous sublayer, for a situation where the fractures are assumed to separate 1 turbulence wall units from each other in the y_2 direction. This observation directly follows from the relation $U^+ \approx y^+$ in the viscous sublayer[2].

[2] Note that enough resolution is required to apply the fracture structure model. In all computations in [23–24], the resolution in the y-direction in the viscous sublayer was selected to be 1 turbulence wall unit.

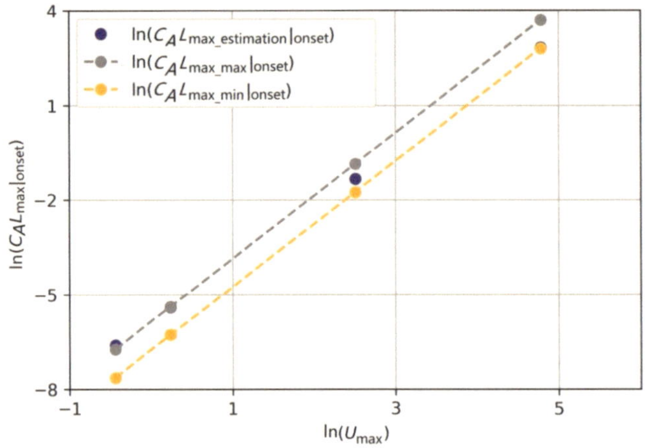

Figure 6.
Graph of $\ln(C_A L_{max}|_{onset})$ *vs.* $\ln(U_{max})$ *for computations of the onset position in the first iteration, according to example case results presented in* ***Table 3***.

If, in the first iteration, it is assumed that the friction velocity at the onset position equals the friction velocity of a corresponding fully developed turbulent pipe flow, then Eq. 2 for onset can be rephrased into:

$$C_A L_{max}|_{onset} = (U^*)^2 \tag{8}$$

For the pipe flow turbulence onset estimations (assuming $\varepsilon/D = 0$), we obtain **Table 3** in the first iteration. Plotting the natural logarithm of the U_{max} on the x-axis, and the natural logarithms of $C_A L_{max}|_{onset}$ on the y-axis, gives **Figure 6**.

The empirical $C_A L_{max}|_{onset}$ numbers occur within the zone between grey dots and orange dots (which corresponds to the Re_x interval between 5×10^5 and 3×10^6). The blue dots represent the estimated $C_A L_{max}|_{onset}$ calculation from Eq. 8, if assuming the friction velocity in the onset position equals the friction velocity of a corresponding fully-developed turbulent pipe flow – which for three of the four cases fall within the onset zone.

Second iteration:

It should again be stressed that the $C_A L_{max}|_{onset}$ estimated in **Table 3** (first iteration), represents the condition in the turbulence onset position, and nowhere else. To

Re_D	U_{max}	U^*	$\frac{\partial U_1}{\partial y_2}\Big\vert_{near\ wall}$ Eq. 4	$C_A L_{max}\vert_{onset}$ Eq. 2	$C_A L_{max}\vert_{onset}$ ($\mathrm{Re}_x = 5 \times 10^5$) Eq. 3	$C_A L_{max}\vert_{onset}$ ($\mathrm{Re}_x = 3 \times 10^6$) Eq. 3
5×10^3	0.64667	0.0368692	90.0223	0.001359	0.000483	0.001183
1×10^4	1.27352	0.0670356	297.6009	0.00449	0.001873	0.004587
1×10^5	12.2575	0.511689	17339.5	0.2618	0.1735	0.4250
1×10^6	119.578	4.11757	1,122,520	16.95	16.51	40.44

Table 3.
Estimating onset position for 14 cm diameter pipe flows, with air flowing at 20°C (first iteration). Data of fully developed pipe flows were taken from [24]. The Centre-line velocity U_{max} *estimated from the log-law relation for smoothness factor 0.0.*

illustrate, the $C_A L_{\max}$ for the corresponding fully developed pipe flow conditions along solid walls at stationary turbulent flow are different than in **Table 3**. According to **Table 1** in [24], the corresponding values are $C_A L_{\max}$ = 0.002103 at $\mathrm{Re}_D = 5 \times 10^3$, $C_A L_{\max}$ = 0.007380 at $\mathrm{Re}_D = 1 \times 10^4$, $C_A L_{\max}$ = 0.5251 at $\mathrm{Re}_D = 1 \times 10^5$, and $C_A L_{\max}$ = 41.39 at $\mathrm{Re}_D = 1 \times 10^6$.

The onset estimation may be improved, by refining the estimations on $C_A L_{\max}|_{\text{onset}}$ and the velocity gradient.

For instance, if the velocity gradient where onset occurs is 50% lower, then it should be noted that adjusting Eq. 4 into $\frac{\partial U_1}{\partial y_2}\Big|_{\text{near wall}} \approx 5\frac{U_{\max}}{\Delta}$, and consequentially adjusting Eqs. 5–7, then a recalculation will result in the following changes in **Table 3** and **Figure 6**: The numbers in the four rightmost columns of **Table 3** will be 50% lower, and **Figure 6** would appear similar, but with a factor 2 difference in plotting all y-axis numbers.

In conclusion, the analysis in Sections 3.2.2–3.2.3 indicates that – despite exact numerical values on flow gradient at the onset position are not known – there appears to exist a single onset position x_{cr} in a developing boundary layer flow.

Also, in terms of $C_A L_{\max}$ numbers, for the guess made in the first iteration arrived at onset of turbulence occurring prior to reaching fully developed turbulent flow conditions in the pipe, which is a necessary result. Also, looking at the $C_A L_{\max}$ numbers for the ranges of Re_x-numbers, all intervals occur prior to the pipe-flow turbulence reaching fully developed flow conditions.

It is not clear, however, if the predicted $C_A L_{\max}$ value at onset for the case $\mathrm{Re}_D = 1 \times 10^6$ is too high, as these numbers are rather close to the $C_A L_{\max}$ at fully developed turbulent conditions, according to data from [24]. A scaling as demonstrated in the second iteration, may be required to adjust the $C_A L_{\max}$ values, when more data becomes available.

3.3 Analysis with classical turbulence theory - failure to predict onset

According to CTT, the viscous sublayer is considered a thin layer where no eddies are present. The irreversible processes assumed in CTT are the near-equilibrium continuum irreversible process of viscous shear flow, resulting in viscous dissipation.

At the onset point, in the vicinity of the solid wall, assuming the flow gradient on the laminar flow side to be $(\partial U_1/\partial y_2)_{\text{laminar}}$, at the turbulence side, the flow gradient is $(\partial U_1/\partial y_2)_{\text{turbulent}}$.

The total kinetic energy dissipation equals the viscous dissipation, which for the laminar flow case can be expressed as $\frac{\mathrm{d(ke)}_{\text{viscous}}}{\mathrm{d}t} = \mu\left(\frac{\partial U_1}{\partial y_2}\right)^2\Big|_{\text{laminar}}$. For the turbulent case, a corresponding expression can be derived.

For a fully developed pipe flow situation, according to CTT, for the same mean flow rate, the friction factor according to the Moody diagram for corresponding fully developed pipe flows is 60–180% larger (depending on wall surface roughness) for the turbulent flows at Re_D = 2300, as compared to the friction factor for laminar flow. There seems to be no overlap of the friction factors in the transitional region within the Moody diagram with the laminar flow friction factor.

Hence, according to Eq. 10 and 12 in [23], the wall shear stress is for the same mean flow rate, will always be different for the turbulent flow case compared to the corresponding laminar flow case, which is also the case at the onset point. Hence, according to CTT, the flow gradient at the wall – directly connected with the wall shear stress – will always be different at all hypothetical onset points, when comparing the flow gradients at the same wall position for situation of turbulent flow or corresponding laminar flow at the same mean flow rate. The same is also the case for kinetic energy dissipation in the vicinity of the wall. Hence, it is not possible to find any common geometrical position according to CTT where the flow gradients are equal for both laminar flow case and turbulence flow case, and simultaneously the local kinetic energy dissipation is equal for both the laminar flow case and corresponding turbulent flow case.

Hence, it is not possible to use the CTT to analytically predict the turbulence onset.

4. Considerations for near-wall flow measurements and simulations

4.1 Cascade theory and 1st law of thermodynamics: Suggesting a low influence of visible eddies on net kinetic energy dissipation

If *relevant processes are better understood, this may prove beneficial when considering apparatus design*.

Consider, e.g., the Cascade theory of Richardson [34] and Kolmogorov [35] and the 1st law of thermodynamics. While so-called "energy budgets" of visible eddies have been presented in CTT literature, however, no obvious connection to the 1st law of thermodynamics has been presented. Hence, is it possible to make a 1st law analysis of the turbulent flow incorporating the cascade theory?

Yes, this may be possible. The results obtained in [24], indicated that the net contribution of the outer layer to the total kinetic energy dissipation was somewhere between 0 and 3.8% (of the total kinetic energy dissipation of the flow) for all cases studied. The net contribution of the log-law region to the total kinetic energy dissipation was somewhere 0–8.7% (of the total kinetic energy dissipation of the flow) for all cases studied. More importantly, the contribution of the kinetic energy dissipation as caused by the visible eddies in these regions appears to be less than these numbers. In Chapter 3.4 in [24] it was argued that the presence of visible eddies connected to the presence of defect zones, i.e. a local zone with reduced influx of work to be consumed by the irreversible process, as these defects could generate the visible eddies. The net effect on the total kinetic energy dissipation in the pipe flow – due to the actions (direct and indirect) of all visible turbulent eddies – was estimated to be within the range of 1–3% (reduced net kinetic energy dissipation) of the net total kinetic energy dissipation of the turbulent flow, cf. estimations in Section 3.4 in [24].

The following arguments can be made that the findings in [24] are in line with the Cascade theory:

- First, the original cascade theory does not state which fundamental model is valid during the cascading of eddies. Only the last step, the conversion of the smallest eddies into viscous dissipation, occurs via Newtons fundamental model. (There is no contradiction here, as stated in [23–24], for the smaller flow gradients, the maximum entropy generation will occur for the Newtons viscous irreversible process, hence within the outer layer, the eddies dying out will eventually reach small gradients, and be converted into viscous dissipation).

- Second, the original cascade theory, in itself, will suggest the net amount of viscous dissipation is caused by the turbulent eddies in a turbulent pipe flow: A

1st law analysis, if strictly following the assumption that larger scale eddies may transform – without any direct or indirect energy interaction with the main time-averaged flow – into smaller-scaled eddies with preserved kinetic energy, this assumption has major implications: The 1st law analysis, due to the downstream translation invariance for a fully developed pipe flow, will show that the work input (or work "injection" as referred to in the Cascade Theory) required to generate the largest-scale eddies is approximately 1–3% of the net kinetic energy of the kinetic energy of the time-averaged flow, obtained from the relation $(\text{kinetic energy})_{\text{largest scales}} = \frac{1}{2}\left((u_1')^2 + (u_2')^2 + (u_3')^2\right)$ (in units J/kg) which is approximately 1–3% of the kinetic energy of the time-averaged flow (the fluctuating component u_1' is the largest fluctuating component in a pipe flow with mean flow in the y_1-direction, approximately 10% of U_{mean}, while the other two fluctuating components are much smaller). Hence, since no kinetic energy dissipation or viscous dissipation is assumed to occur in the cascading process, and no kinetic energy is transferred from the mean flow to the smaller-scale eddies, it is a physical necessity that the last step which converts kinetic energy into viscous dissipation represents the same 1–3% of the net input kinetic energy, i.e. the net contribution of all visible/cascading eddies to the dissipation of total turbulent flow is only some 1–3%. (We keep in mind that work input in the form of maintaining a steady work input per unit time, while dissipation occurs across a certain volume – which is why units for kinetic energy of time-averaged flow and largest eddies differ from the units of dissipation, while the specific percentage number range 1–3% remains the same. This same 1–3% influence of the defect zones (generated by visible eddies) of the total kinetic energy dissipation of the total flow was also obtained in the results estimated in [24]. (Most important to note: While cascade theory correctly describes 100% of the viscous dissipation to be caused by the turbulent eddies at the smallest Kolmogorov scales, it does nowhere claim this viscous dissipation to represent the 100% kinetic energy dissipation of the entire turbulent flow).

4.2 Planning experiments within viscous sublayer and buffer region

If designing experiments on turbulent boundary-layer flows, the insights gained by the novel approach (if the experimentalist would wish to consider the behaviour predicted in this paper, and in [23–24]), the following recommendations are made:

- Consider zones of high net kinetic energy dissipation as more important to experimentally analyse, as compared to low-kinetic energy dissipation zones. For instance, the viscous sublayer has the highest kinetic energy dissipation, and is for this reason the zone where turbulence onset triggers.

- Consider experimental mean velocity flow meters based on the Hot-Wire technique as dependable, except in the vicinity of a solid wall (where heat losses to the wall may disturb the measurement). The mean flow profile, after non-dimensional scaling, can be connected to L and δ in various zones across the boundary layer.

- Be aware of the potential differences in onset/triggering behaviour, where the presence of a nano-scale particle may appear to have negligible disturbance within a continuum viscous process flow gradient, while, in contrast, may very well result in unexpected behaviour within a far-from-equilibrium process condition. (A surface holographic interference analysis apparatus was developed by [36], however they *failed to obtain spatial resolution when testing within a viscous sublayer, for reasons unknown* [37]). Most turbulence scientists, for instance, are familiar with the ease at which a natural-convection smoke fume from a stationary cigarette at ambient indoor conditions onsets into turbulence. The following question should be posed: Does the smoke particles contribute to the onset of turbulence in this case?

- To allow basic analysis with the novel approach, initial data extracted from a Moody diagram or corresponding Colebrook equation can be used, for analysis in pipe flows. For instance, all computations in [23–24] utilised the Colebrook equation, and the results from [24] was also applied in this paper, to estimate onset condition along a developing-boundary layer flow along a solid wall. The experimental tools originally used to create the Moody diagram (where Colebrook equation is a mathematical formula providing Moody data within 1.5% accuracy) where recorders of pressure drop, recording devices for the mean flow rate, as well as a scale for the representation of the wall surface roughness, expressed in dimensionless form relative roughness ε/D. Also, experimental tools to record the viscosity of the Newtonian fluid is necessary, to compute the relevant Reynolds number. The Moody chart was obtained with data obtained at or around ambient temperatures.

- To consider experiments for model improvements, one approach is to consider better understanding of $L_{\max}$ and δ within the viscous sublayer. Visualisation techniques and the introduction of dye or gas bubbles could be made within the viscous sublayer, to estimate the nature and size scales of $L_{\max}$ and δ. Some near-wall visualisation techniques may be used, for this purpose. Through dye injection or injection of hydrogen bubbles (by electrolysis), one has with high-speed video motion photography observed phenomena such as "burst-sweep cycles," which include the description of motion of "ribbon-like" patterns of fluid within the viscous sublayer, resulting in the variation of the thickness of the viscous sublayer with time. In [23–24], a discussion on the "overproduction" of "flakes" describes the release of far-from-equilibrium zones which release and flow downstream (moving downstream with a significantly lower speed, as compared to the local speed of the fluid flow itself).

- In [23–24], a discussion on the shape of these flakes, which have a width K and length L (as the streaks align with the flow), which might connect with the length of the low-speed streaks (note that the flakes themselves may be moving with low speed or being stationary). The upwards movement experimentally found in these streaks is due to downward flow in high-speed areas (to fulfil the conservation of mass principle).

- In experiments, it has been found that the streaks shift inconsistently, and appear and disappear over time. In connection with the proposed novel approach, this could suggest that the flakes are occasionally not in contact with each other in the transverse-direction (y_3-direction) of the fluid flow direction (y_1-direction). The widths between these streaks suggest $K\approx100+$ (turbulence wall units), while the length of these streaks suggest $L\approx1000+$ (turbulence wall units) in length. (The Re numbers do not influence these numbers.)

- When studying the streaks, it is found that similar traces of streaks may be found up through the buffer- and log-law regions, maintaining the same width between the streaks, but with reduced length [38] – also in excellent agreement with novel approach. However, due to the visible eddy mixing, the presence of streaks weakens when moving further away from the wall.

- However, for the study of the effects of mixing, or downstream development of a turbulent wall-bounded flow boundary-layer thickness, the understanding of the rotation of the visible eddies do play an important role. This is particularly important for the study of e.g. heat transfer behaviour.

- One should not rule out the possibility of recording sound waves generated from the turbulent flow, to perform an analysis. In some reports [39], sound energy maxima are recorded at specific frequencies [39]. Is there any possibility to connect such maximums to the novel approach theory?

4.3 Planning of more detailed simulations accounting for an overall active far-from-equilibrium mechanism

It was estimated in [24] and in Section 4.1 that ignoring the transient simulation of swirls and eddies resulted in errors in kinetic energy dissipation of around 1–3%. The results of the steady-state simulations appear accurate enough to allow the computations of turbulence onset for fully-developed pipe flows, as well as for a developing-flow boundary layers – as presented in Section 3.

However, it should be mentioned that some aspects of a more detailed simulation or modelling – if one would wish to simulate various details of the near-wall flow behaviour – is possible, however, a bit more complex as compared to the complexity in existing CFD flow solvers. (A multiphase CFD flow solver may need to be used.) This aspect touches on the drawbacks with the novel approach: Consider, for instance, how to simulate the individual near-wall fracture cells. These may be active and be located at stationary positions. But these may also suddenly start to act in a non-stationary manner, by moving downstream. To further complicate the picture, if these cells move downstream, they will move at a much lower velocity as compared to the speed of a local fluid element.

It is important to distinguish the fracture cells behaviour from the fluid flow behaviour. If assuming that the behaviour of the fracture cells is controlled by an overall MEP process behaviour – which admittedly is not accurately known – provide formidable challenges for a simulation. Also, the movement of defects within the flow (directly impacting the vorticity or visible eddy behaviour), or if these are stationary or not (for instance, more stationary within the buffer region, while separated saturated zones are moving as small “islands” along the flow within the outer region of a turbulent boundary layer) are not well known.

Admittedly, existing far-from-equilibrium theory may not yet give accurate answers to all questions raised, but possibly provide some guidance on the turbulent flow behaviour, which connects to more complex modelling issues such as typical anisotropic behaviour of visible eddies [32] and memory effects [40] (or “coherence” in the turbulent fluid flow), as well as e.g. near-wall sweep-burst behaviour (cf. discussion of leakage of *Residual-Thermodynamic Process* zones in viscous sublayer, cf. [23–24]), and approximately explain the location and behaviour of near-wall turbulent streaks, the size of these, the spacing of these, and why they occasionally vanish.

On a positive note, however, this "unknown behaviour" of MEP could be investigated in numerical simulations by – for instance – testing various injections of defects or defect clouds into various positions in the flows and observing the outcome. If a CFD flow solver is designed to be capable of simulating fracture cell movements and allowing for the simulation of visible eddies by injection of defect or defect clouds, then this may provide a great opportunity to investigate what type of MEP input would give a truly realistic turbulence simulation behaviour output. In this way, the observation of the above-mentioned various experimental phenomena near walls in the different near-wall zones, may become possible to simulate. For instance, phenomena such as apparent turbulent intensities (2nd-order statistics) in the flow may be simulated and evaluated.

5. Discussion and conclusions

A proposed novel approach on the modelling of turbulent flows, offers possibilities of additional understanding on several aspects of turbulent flows.

On the question of how the mechanism of onset of turbulence works; A discussion is presented, arguing for the comparison of flow conditions at a hypothetical onset point in the vicinity of a wall. By applying the novel approach model for the turbulence modelling, the onset Re numbers for fully developed pipe flows (computed in [24]) and the onset Re numbers for a developing-flow boundary layer along a wall can be computed. The onset estimations arrive at excellent agreement between modelling results and corresponding experiments. By applying the CTT approach, it is here demonstrated as well that no onset Reynolds numbers can be computed.

On the question of benefits obtained with the proposed novel approach, one may argue for a good connection of the novel approach with the original boundary-layer theory introduced by Prandtl, with possibilities to apply original dimensionless scaling concepts such as U^+, U^* and turbulent wall dimension units. The parameter $C_A L_{\mathrm{max}}$, which can be extracted by model-fitting a couple of points in a Moody-diagram, provides an exponential correlation between $C_A L_{\mathrm{max}}$ and U_{mean} (for data extracted at a specific wall surface roughness ε/D), for all Reynolds numbers, cf. Eqs. 1 and 2 in [24], a correlation which also works when extrapolating far outside the points used for model-fitting, into a Reynolds numbers region where onset analysis can be successfully applied. (Hence, it is not required to fit model coefficients at actual onset flow conditions to predict onset.) One should also note that the novel approach draw attention to the problem of turbulence in an unprejudiced manner, with the core aim of analysing the large-scale behaviour with as few assumptions and simplifying assumptions as possible. While the specific assumptions and simplifying assumptions made for CTT can be criticised, it is also in this context worthwhile to mention the criticism presented in [41]: "So confident was the turbulence community in its beliefs that virtually no one even bothered to measure the skin friction, and it was simply inferred from fitting the log profile to a few points near the wall, usually for values of y+ between 30 and 100. In fact the 'log'-based ideas were so well-accepted that it seemed to bother only a few that real shear stress measurements (both momentum integral and direct) differed consistently and repeatably from these inferred results." The results presented in **Figure 2** may explain the reasons for consistent and repeatable erratic prediction of the real wall shear stress.

Acknowledgements

Hot Disk AB (Sweden) supported this work. Special thanks to Prof. L. Löfdahl at Dept. of Mechanics and Maritime Sciences, Chalmers Univ. of Technology, D.Sc. S.E. Gustafsson at Dept. Physics, Chalmers Univ. of Technology and Dr. H. Otterberg at University of Gothenburg. Also, thanks to Dr. B. Mihiretie, Mr. A. Sizov and Ms. B. Lee at Hot Disk AB (Sweden), for assistance in preparing this manuscript.

Conflict of interest

The author declares no conflict of interest.

Nomenclature

A	area (m^2)
C_A	model constant ($m\ s^{-2}$)
C_B	model constant (s)
D	diameter of pipe interior (m)
$\frac{d(ke)_{res}}{dt}$	kinetic energy dissipation rate per unit volume, due to residual process ($W\ m^{-3}$)
$\frac{d(ke)_{viscous}}{dt}$	viscous dissipation rate for viscous laminar flows of Newtonian fluids per unit volume ($W\ m^{-3}$)
F	force (N)
K	flake width (in y_3-direction) (m)
L	slip length of flake (in y_1-direction), cf. [23] (m)
R	radius of pipe interior (m)
Re	Reynolds number (–)
S	entropy ($J\ K^{-1}$)
U	Velocity ($m\ s^{-1}$)
u'	Fluctuating velocity (after Reynolds decomposition) ($m\ s^{-1}$)
U^*	Friction velocity ($m\ s^{-1}$)
U_{slip}	slip velocity ($m\ s^{-1}$)
x	distance from front edge of plate (m)
X	extensive thermodynamic flow
y	vertical distance (from the plate surface) (m)
z	longitudinal-direction distance (from a hypothetical reference point or line, e.g., centreline of plate) (m).
y_1, y_2, y_3	Cartesian co-ordinates (m)

Greek letters

Δ	boundary layer thickness (m)
δ	gap width between slip layers, also referred to as thickness of flake fracture, or resolution parameter (m)
ε	surface roughness (m)
μ	dynamic viscosity of the Newtonian fluid ($kg\ m^{-1}\ s^{-1}$)
ν	kinematic viscosity of the Newtonian fluid ($m^2\ s^{-1}$)

ξ	model constant, introduced in Eq. (3) (–)
ρ	density (kg m^{-3})
$-\rho\overline{u_1 u_2}$	Reynolds stress (N m^{-2})
τ	shear stress (N m^{-2})

Subscripts

cr	(critical)
D	inner diameter-based
max	(maximum)
mean	(average)
near wall	vicinity of wall
onset	condition at the onset position
res	for residual sub-process
x	flat plate-direction

Special notations

$(\cdot)^+$	dimensionless scaling (traditional), cf. **Figure 2** in [24]
$\mathrm{d}(\cdot)$	differential [e.g., $\mathrm{d_i S_{res}}$ represents the differential entropy change due to residual process (J K^{-1})]

Author details

Mattias K. Gustavsson
Hot Disk AB, Gothenburg, Sweden

*Address all correspondence to: mattias.gustavsson@hotdiskinstruments.com

References

[1] Phillips L. Turbulence, the Oldest Unsolved Problem in Physics. 2018. Available online: https://arstechnica.com/science/2018/10/turbulence-the-oldest-unsolved-problem-in-physics/

[2] White FM. Fluid Mechanics. 2nd ed. Blacklick, Ohio, USA: McGraw-Hill Book Company; 1986

[3] Panton RL. Incompressible Flow. 4th ed. New York, USA: John Wiley & Sons; 1984

[4] Tritton DJ. Physical Fluid Dynamics. 2nd ed. New York, USA: Oxford University Press; 1988. p. 519

[5] Tennekes H, Lumley JL. A First Course in Turbulence. Cambridge, Massachusetts, USA, and London, England: MIT Press; 1972

[6] Davidson PA. Turbulence: An Introduction for Scientists and Engineers. New York, USA: Oxford University Press; 2004

[7] Davidson PA, Kaneda Y, Moffatt K, Sreenivasan KR, editors. A Voyage through Turbulence. Cambridge, UK: Cambridge University Press; 2011

[8] Monin AS, Yaglom AM. Statistical Fluid Mechanics. Vol. 1. Cambridge, MA: MIT Press; 1971. p. 769

[9] Robinson SK. Coherent motions in the turbulent boundary layer. Annual Review of Fluid Mechanics. 1991;**23**:601-639

[10] Smith CR. Coherent flow structures in smooth-wall turbulent boundary layers: Facts, mechanisms and speculation. In: Ashworth PJ, Bennett SJ, Best JL, McLelland SJ, editors. Coherent Flow Structures in Open Channels. Chichester, UK: Wiley; 1996. pp. 1-39

[11] Corino ER, Brodkey RS. A visual investigation of the wall region in turbulent flow. Journal of Fluid Mechanics. 1969;**37**:1-30

[12] Offen GR, Kline SJ. Combined dye-streak and hydrogen-bubble visual observations of a turbulent boundary layer. Journal of Fluid Mechanics. 1974;**62**:223-239

[13] Offen GR, Kline SJ. A proposed model of the bursting process in turbulent boundary layers. Journal of Fluid Mechanics. 1975;**70**:209-228

[14] Harun Z, Lofty ER. Generation, Evolution, and Characterization of Turbulence Coherent Structures. London, UK: IntechOpen; 2018. DOI: 10.5772/intechopen.76854

[15] Demirel Y. Nonequilibrium Thermodynamics: Transport and Rate Processes in Physical, Chemical and Biological Systems. 3rd ed. Amsterdam, The Netherlands: Elsevier; 2014

[16] Herrmann B, Oswald P, Semaan R, Brunton SL. Modeling synchronization in forced turbulent oscillator flows. Communications on Physics. 2020;**3**:195. DOI: 10.1038/s42005-020-00466-3

[17] Moin P, Mahesh K. Direct numerical simulation: A tool in turbulence research. Annual review of fluid mechanics. 1998;**30**:539-578. DOI: 10.1146/annurev.fluid.30.1.539

[18] Liu C, Lu P, Chen L, Yan Y. New Theories on Boundary Layer Transition and Turbulence Formation. Modelling and Simulation in Engineering. Cairo,

Egypt: Hindawi Publishing Corporation; 2012; Article ID 619419. DOI: 10.1155/2012/619419

[19] Bose R, Durbin PA. Transition to turbulence by interaction of free-stream and discrete mode perturbations. Physics of fluids. 2016;**28**:114105. DOI: 10.1063/1.4966978

[20] Ducros F, Comte P, Lesieur M. Large-eddy simulation of transition to turbulence in a boundary layer developing spatially over a flat plate. Journal of fluid mechanics. 1996;**326**: 1-36. DOI: 10.1017/S002211 2096008221

[21] Launder BE, Spalding DB. The numerical computation of turbulent flows. Computer methods in applied mechanics and engineering. 1974;**3**: 269-289. DOI: 10.1016/0045-7825(74) 90029-2

[22] Hinze JO. Turbulence – An Introduction to its Mechanism and Theory. U.S.A.: McGraw-Hill; 1959

[23] Gustavsson M. A residual thermodynamic analysis of turbulence – Part 1: Theory. International Journal of Thermodynamics. 2022;**25**:50-62. DOI: 10.5541/ijot.1017342

[24] Gustavsson M. A residual thermodynamic analysis of turbulence – Part 2: Pipe flow computations and further development of theory. International Journal of Thermodynamics. 2022;**25**:64-75. DOI: oi.org/10.5541/ijot.1017374

[25] Jaeger HM, Liu AJ. Far-from-equilibrium physics: An overview. arXiv. 2010. p. 1-23. DOI: 10.48550/arXiv.1009.4874. Eprint arXiv: 1009.4874; EPub date: September 2010

[26] Kondepudi D, Prigogine I. Modern Thermodynamics: From Heat Engines to Dissipative Structures. Chichester, UK: Wiley; 1998

[27] Gustavsson M. Residual thermodynamics: A framework for analysis of non-linear irreversible processes. International Journal of Thermodynamics. 2012;**15**:69-82. DOI: 10.5541/ijot.346

[28] Kleidon A, Malhi Y, Cox PM. Maximum entropy production in environmental and ecological systems. Philosophical Transactions of the Royal Society B. 2010;**365**:1297-1302. DOI: 10.1098/rstb.2010.0018

[29] Finnie I, Kabil YH. On the formation of surface ripples during erosion. Wear. 1965;**8**:60-69. DOI: 10.1016/0043-1648 (65)90251-6

[30] Emmons HW. The laminar-turbulent transition in a boundary layer – Part I. J. Aero. Sci. 1951;**18**:490-498. DOI: 10.2514/8.2010

[31] Emmons HW, Bryson AE. The laminar-turbulent transition in a boundary layer (part II). In: Proceedings of the first U.S. National Congress of Applied Mechanics (June 1951, Chicago, Illinois) A.S.M.E., 1952. pp. 859-868

[32] Davis SH, Lumley JL. Frontiers in Fluid Mechanics. A Collection of Research Papers Written in Commemoration of the 65th Birthday of Stanley Corrsin. Berlin Heidelberg, Germany: Springer Verlag; 1985

[33] Jonáš P. On the turbulent spot and calmed region. In: Proceeding of Engineering Mechanics 2007, National Conference with International Participation. Prague, Czech Republic: Institute of Thermomechanics, Academy of Sciences of the Czech Republic; May 14-17, 2007

[34] Richardson LF. Weather Prediction by Numerical Processes. Boston: Cambridge University Press; 1922. p. 66 (ISBN 9780511618291.)

[35] Kolmogorov AN. Local structure of turbulence in an incompressible fluid at very high Reynolds numbers. Doklady Akademii Nauk SSSR. 1941;**31**:99-101

[36] Abrantes JK. Holographic Particle Image Velocimetry for Wall Turbulence Measurements, Ph.D. thesis,. Lille, France: Ecole Centrale de Lille; 2012

[37] Abrantes JK. Private communication. 2023

[38] Wang W, Pan C, Wang J. Wall-normal variation of spanwise streak spacing in turbulent boundary layer with low-to-moderate Reynolds number. Entropy. 2019;**21**:24

[39] Succi GP. The Interaction of Sound with Turbulent Flow, Gas Turbine & Plasma Dynamics Laboratory Report no. 140. Cambridge, MA, U.S.A.: MIT; 1977

[40] Ross D, Robertson JM. Shear stress in a turbulent boundary layer. Journal of Applied Physics. 1950;**21**:557-561. DOI: 10.1063/1.1699706

[41] George WK. Recent advancements toward the understanding of turbulent boundary layers. In: Proceedings of the American Institute of Aeronautics and Astronautics. Reston VA, USA: American Institute of Aeronautics and Astronautics; 2005-4669

Section 2

Drag Reduction Initiatives Using Active Control

Chapter 3

On Drag Reduction by Spanwise Wall Oscillation in Compressible Turbulent Channel Flow

Wenjun Yuan, Dongming Chen and Haifeng Zhang

Abstract

In this chapter, we summarize our progress in the investigation of drag reduction (DR) by spanwise wall oscillation (SWO). Reliable direct numerical simulation (DNS) models have been established for compressible turbulent channel flow. A large amount of 39.3% drag reduction has been achieved for supersonic compressible flows. It is found that the compressible effect is modest with respect to the turbulent effect for the skin-fiction in compressible turbulent channel flows. The reduced DR is mainly because of the decreasing turbulent contribution, and the related small compressible term also slightly decreases with the increase $W_m{}^+$. More DNS cases with different maximum wall velocities, oscillation periods, and flow Reynold numbers for compressible cases should be analyzed. The optimal combination with the highest drag reduction efficiency has significant importance on real applications, which deserves to be studied in detail to characterize the underlying mechanisms.

Keywords: compressible flow, wall turbulence, drag reduction, spanwise wall oscillation, direct numerical simulation

1. Introduction

Flow control aiming at reducing turbulent skin friction is a challenging theoretical and technological problem, which is of great importance for energy saving, aircraft designation and pipeline transportation. Knowledge of the dynamics of near-wall turbulent structures can provide a thorough understanding of the underlying physics and has been developed into various drag reduction control methods in the past decades [1]. However, typical passive controls without energy input, such as riblets [2], random wall roughness [3], polymers [4], superhydrophobic surfaces [5], and dimples [6], are hard to continuously sustain in practical applications [7]. Spanwise wall oscillation (SWO), as an efficient active control technique with significant drag reduction efficiency, has received increasing attention in recent years.

Among the intriguing research work, Jung et al. [8] and Akhavan et al. [9] were the first to adopt direct numerical simulation (DNS) to investigate the suppression of turbulence when a spanwise oscillating wall was imposed. In their simulations, as demonstrated by Kim et al. [10], the periodic boundary conditions were adopted in homogeneous directions. At the channel walls, the usual no-slip and no-penetration conditions were applied. The wall boundary condition for the spanwise component of velocity was:

$$W = W_m \sin(\omega t) = W_m \sin\left(\frac{2\pi}{T}t\right) \tag{1}$$

i.e. the two walls move in phase with a spanwise velocity W which is a sinusoidal function of time t with prescribed amplitude ${W_m}^+$ and oscillation frequency ω (or period T). They observed that 40% skin-friction drag could be achieved in a numerical channel at $Re_\tau = 180$, and the optimal wall oscillation period was found to be ${T_{opt}}^+ \approx 100$. Following this work, Baron and Quadrio [11] further considered both energetic costs and benefits of this active technique, and showed that around 10% net energy saving was possible by spanwise wall oscillation, which was a similar level to the passive drag reduction controls, such as riblets. This drag reduction performance by the spanwise wall oscillation was confirmed experimentally by Laadhari et al. [12], Di Cicca et al. [13], and Ricco [14] for the flat plate wall. In the present project, we also impose these no-slip time-dependent boundary conditions at the walls to study the drag reduction by SWO in turbulent channel flow.

Most studies on drag reduction by SWO have focused on the two control parameters, i.e., the maximum wall velocity ${W_m}^+$ and the oscillation frequency ω^+, and explored a scaling law of the drag reduction. It was concluded that the drag reduction value was not simply scaled by the maximum wall displacement ${D_m}^+$ nor the oscillation frequency ω^+. Choi and Graham [15] experimentally studied the drag reduction in a circular pipe at two Reynolds numbers, $Re_\tau = 650$ and 1000, and found that the maximum wall velocity ${W_m}^+$ gave a better scaling than the oscillation frequency ω^+. Choi et al. [16] proposed a combined number V_c with a thickness l^+, an acceleration rate a^+, the maximum wall velocity ${W_m}^+$, and the Reynolds number Re_τ: ${V_c}^+ = a^+ l^+ / {W_m}^+ Re_\tau^{-0.2}$, and the authors found that the drag reduction was scaled in a quadratic form of ${V_c}^+$, i.e., $\mathrm{DR} = 1000{V_c}^+ + 50\ {V_c}^+$. Similarly, Quadrio and Ricco [17] proposed a combined parameter $S^+ = a^+ l^+ / {W_m}^+$ without the consideration of the Reynolds number effect. The authors found that the drag reduction was scaled linearly well with S^+, when the oscillation period was small, i.e., $T^+ < 150$, which was also the accurate prediction regime of the model-based approach by Moarref and Jovanović [18]. Ricco and Quadrio [19] further explored the parameter space T^+ - ${W_m}^+$ and T^+ - ${D_m}^+$ for the drag reduction region and also for the net energy saving condition at $T^+ < 150$, where the linear scaling correlation is held.

In spite of their various captivating features, the drag reduction mechanism of spanwise oscillations has been partially understood [1]. **Table 1** presents different models coexisted in the literature. Akhavan et al. [9] and Baron and Quadrio [11] pointed out that the spanwise wall oscillation generated the Stokes layer, and it shifted the position of the low-speed streaks relative to the quasi-streamwise vortices. Dhanak and Si [27] further considered a single coherent quasi-streamwise vortex dipole in a spanwise oscillation flow and argued that the spanwise wall oscillation deformed the quasi-streamwise vortices and increased the mixing of the high-speed and low-speed streaks, resulting in the drag reduction. Nevertheless, Galionis and Hall [28] theoretically studied the unstable Görtler vortex on a concave surface subjected to the spanwise wall oscillation. The growth rate of the most amplified Görtler vortex was found to be significantly reduced. Negi et al. [29] used a localized volume forcing to

Source	Mechanism
Choi and Graham [15]	Negative spanwise vorticity is created in both positive and negative movement of spanwise wall oscillation, reducing velocity gradient in the near wall region.
Xu and Huang [20]	The global turbulence suppression is caused by the sustained attenuation of the pressure strain term in the turbulent budget.
Duggleby et al. [21]	The coherent vorticity structures are pushed away from the wall into higher speed flow, causing a shorter time interaction between the propagating wave modes and the roll modes, thus less Reynolds shear stress production.
Ricco et al. [22]	Turbulent dissipation rate is enhanced during the transient process, which leads to drag reduction in the new quasi equilibrium state.
Touber and Leschziner [23]	The unsteady cross-flow straining causes major spanwise distortions in the streaks, and the reduction in wall normal and shear stresses.
Agostini et al. [24]	The drag reduction process is linked to the change in the Stokes strain in the upper region of the viscous sublayer where streaks are strong.
Yakeno et al. [25]	Spanwise wall oscillation suppresses the near-wall streamwise vortices rotating in the opposite direction, and also tilt the structures into spanwise direction.
Gatti and Quadrio [26]	The large structures, which contribute to turbulent friction via the Reynolds shear stress produced in the outer region, do not directly jeopardize the drag reducing action-based spanwise forcing.

Table 1.
Drag reduction mechanisms by spanwise wall oscillation in literature.

generate the low-speed streaks in a laminar boundary layer and studied the interaction between the low-speed streaks and the spanwise wall oscillation. The authors found that the drag reduction values had a better correlation with the wall-normal velocity fluctuations.

DNS provides an opportunity to critically examine the coherent structures in the turbulent flow due to its ability to simulate highly accurate 3D flow fields. With the visualization technique, Blesbois et al. [30] found that in a spanwise oscillating turbulent boundary layer, there were infinitely long structures with certain angles to the mean flow, and the angle and amplitude of the structures jumped suddenly at a certain instant during the oscillation period, which was consistent with the conditioned streaks angle in the DNS by Touber and Leschziner [23]. Yakeno et al. [25] performed the ensemble average of the quasi-streamwise vortices and the associated quadrant events, and argued that the drag reduction for the cases with small oscillation periods was due to the suspension of the Q2 event; while the drag increase for the cases with large oscillation periods was due to the enhancement of the Q4 event. Touber and Leschziner [23] compared the turbulent statistics between the spanwise wall oscillation case and the no control case in much detail and pointed out that the origin of the statistics changes was the spanwise distortion of the near wall streaks. This mechanism was further supported by Agostini et al. [24] at a higher Reynolds number, and the drag reduction by the spanwise wall oscillation was due to the increase of turbulent enstrophy and dissipation in the transient process [22]. Recently, Gatti and Quadrio [26] and Agostini and Leschziner [31] focused on the impact of footprints of large-scale outer structures on the near-wall layer in the presence of drag-reducing spanwise wall motion. It was noted that the large structures, which contribute to turbulent friction via the Reynolds shear stress produced in the outer region [32], do not directly jeopardize the drag-reducing action of the wall-based spanwise forcing. However, the physical mechanisms underlying the SWO are still not clear, with unanswered questions such as how the oscillation affects the near-wall dynamics and how the resultant self-sustaining cycle impacts the suppression of streamwise vortex generation and hence reduction of drag.

It has been pointed out that streamwise vortices play an important role in all self-sustaining mechanisms of near-wall turbulence [33]. There are different explanations for how these eddies are created and maintained. For example, Hamilton et al. [34] reported that the nonlinear advection term is mainly responsible for flow direction vortices due to flow direction disturbances, while Schoppa and Hussain [35] claimed that the vortex stretching term is mainly responsible for flow direction vortices. However, streamwise vortices are formed and maintained autonomously by a self-sustaining process, which involves wall-layer streaks and instabilities associated with them [36]. In addition, the current literature on SWO largely investigated incompressible flow in turbulent boundary layer, channel, and pipe flows. It was illustrated that the local self-sustaining process also existed for the turbulence in the inner region of compressible turbulent boundary layers for different Mach numbers [33]. Relatively few investigations have been carried out on high-speed compressible flows. Fang et al. [37, 38] and Ni et al. [39] employed large eddy simulation (LES) and DNS to study SWO with compressible turbulent channel flow at $Ma = 0.5$ and a supersonic turbulent boundary layer at $Ma = 2.9$, respectively. In these studies, they found that SWO simultaneously suppresses momentum transportation and heat transportation in a consistent manner. Previous studies of compressible wall-bounded flows have led to the widely accepted notion that the main effect of flow compressibility resides in the incurred mean density variations [33].

Recently, Yao and Hussain [40, 41] presented a new dynamical vortex definition for compressible flows, which is derived directly from the compressible momentum (Navier–Stokes) equation and identifies the coherent local pressure minimum in a plane. Based on this work, they further performed DNS for isothermal channel flow with SWO to investigate subsonic and supersonic flows at bulk Mach numbers of 0.8 and 1.5, respectively [42]. Their results suggested that significant drag reduction can still be achieved via SWO even when the compressibility effect is considered. Therefore, an attempt to extend the research work to account for compressibility effects at transitional Mach numbers and reveal the detailed drag reduction mechanism by SWO in these flows urgently needs to be addressed.

2. Research progress and results

2.1 Establishment of a reliable DNS model for turbulent channel flow

As shown in **Figure 1**, the turbulent channel flow consists of two infinitely large parallel plates which are at a finite distance of $2h$ apart. Periodic boundary conditions are applied in the streamwise (x) and spanwise (z) directions, respectively. The conservation equations for solving compressible flow problems include continuity, momentum, and energy equations. For a Newtonian fluid behaving as an ideal gas, these equations can be expressed as,

$$\frac{\partial \rho}{\partial t} + \frac{\partial \rho u_i}{\partial x_i} = 0, \tag{2}$$

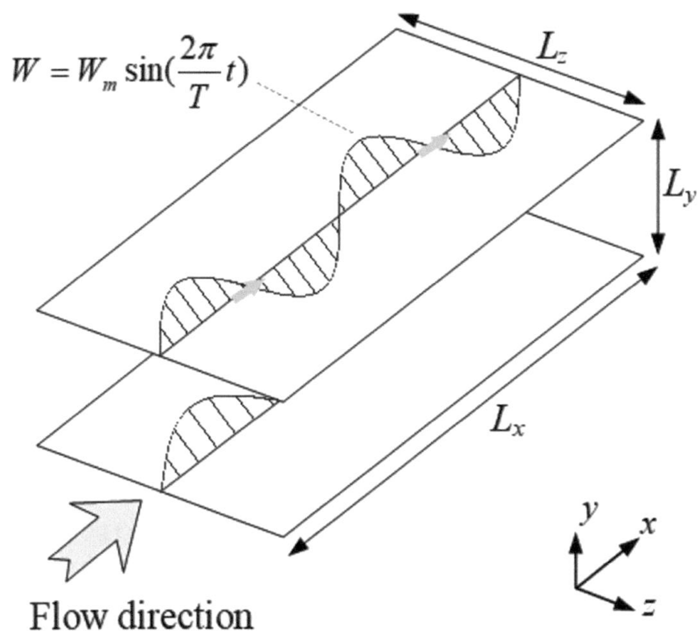

Figure 1.
The schematic of turbulent channel flow with spanwise wall oscillation.

$$\frac{\partial \rho u_i}{\partial t} + \frac{\partial \rho u_i u_j}{\partial x_j} = -\frac{\partial p}{\partial x_i} + \frac{\partial \sigma_{ij}}{\partial x_j} + f\sigma_{i1}, \tag{3}$$

$$\frac{\partial \rho E}{\partial t} + \frac{\partial \rho u_j H}{\partial x_j} = -\frac{\partial q_j}{\partial x_j} + \frac{\partial \sigma_{ij} u_i}{\partial x_j} + f u_1, \tag{4}$$

where u_i is the velocity component in the ith direction, ρ the density, p the pressure, $E = c_v T + u_i u_i/2$ the total energy per unit mass, and $H = E + p/\rho$ is the total enthalpy. $\gamma = c_p/c_v = 1.4$ is the specific heat ratio. q_j and σ_{ij} are the components of the heat flux vector and the viscous stress tensor, respectively,

$$\sigma_{ij} = \mu\left(\frac{\partial u_i}{\partial x_j} + \frac{\partial u_j}{\partial x_i} - \frac{2}{3}\frac{\partial u_k}{x_k}\delta_{ij}\right), \tag{5}$$

$$q_j = -\kappa\frac{\partial T}{\partial x_j}, \tag{6}$$

where the dependence of the viscosity coefficient on temperature is accounted for through Sutherland's law and $k = c_p \mu / P_r$ is the thermal conductivity, with $P_r = 0.72$. The forcing term f in Eq. (3) is evaluated at each time step in order to discretely enforce constant mass-flow-rate in time, and the corresponding power spent is added to the right-hand-side of the total energy equation.

As shown in **Table 2**, the physical parameters used in our DNS are presented. The computational domain is $12h \times 2h \times 6h$ with $320 \times 160 \times 210$grid points in the streamwise (x), wall-normal (y), and spanwise (z) directions, respectively. The grid spacing is constant in the wall-parallel directions, and an error-function mapping is

	R_b	Re_τ	Domain size $L_x \times L_x \times L_z$	Grid points $n_x \times n_x \times n_z$	Δx^+	Δy^+	Δz^+
Coleman et al. [43]	3000	222	$4\pi h \times 2h \times 4/3\pi h$	$144 \times 119 \times 80$	19	$0.1 \sim 8.0$	12
Morinishi et al. [44]	3000	218	$4\pi h \times 2h \times 4/3\pi h$	$120 \times 180 \times 120$	23	$0.36 \sim 5.1$	7.6
Foysi et al. [45]	3000	221	$4\pi h \times 2h \times 4/3\pi h$	$192 \times 150 \times 128$	14.46	$0.84 \sim 5.02$	7.23
Sun et al. [46]	3000	220	$4\pi h \times 2h \times 2\pi h$	$160 \times 160 \times 160$	17.3	$0.05 \sim 5.5$	8.6
Modesti and Pirozzoli [47]	3000	215	$6\pi h \times 2h \times 2\pi h$	$512 \times 128 \times 256$	8.0	-	5.9
Present DNS	3000	200	$12h \times 2h \times 6h$	$320 \times 160 \times 210$	7.46	$0.19 \sim 4.1$	5.7

Table 2.
Setup of compressible channel DNS for supersonic flows at bulk Mach number $M_b = 1.5$, with the friction Mach number $M_\tau = u_\tau / c_\omega = 0.079$ and the heat flux coefficient $B_q = q_\omega / (\rho_\omega C_p u_\omega) = 0.048$.

used to cluster grid points towards the walls. This domain size is compared to previous studies for compressible turbulent channel flow [44], and is also dictated by the need to accommodate the large eddies which become energetically relevant at sufficiently high Reynolds numbers [48, 49]. At a fixed bulk Reynolds number $Re_b = h\rho_b u_b / \mu_w = 3000$, we have carried out DNS of subsonic and supersonic flows at bulk Mach numbers $Ma_b = u_b / c_w = 0.8$ and 1.5, respectively. Here, u_b and ρ_b are the bulk channel velocity and density, and c_w is the speed of sound evaluated at the wall.

In addition, it is noted that the presence of the walls in y direction may lead to failure in the conservation of mass in co-located flow solvers [43, 50], which is typically fixed by adding a source term to the continuity equations at the wall nodes in such a way that the integrated density remains discretely constant in time. In this work, we prefer to stagger the first node off the wall in such a way that the latter coincides with an intermediate node, where the convective fluxes are identically zero. Hence, correct telescoping of the numerical fluxes is guaranteed, and no net mass variation can occur. A further benefit of this approach is that, for a given distance of the first grid point from the wall, the maximum allowable time step associated with the vertical grid spacing is doubled. In order to maximize the spectral resolution in the streamwise direction, all simulations are performed in a convective frame of reference [51], in which the bulk velocity is zero. All computations are initiated with a parabolic velocity profile, superposed random perturbations, and uniform values of density and temperature.

Foysi et al. [45] studied the compressible turbulent channel by using DNS. They also observed a change in the Reynolds normal stresses at high Mach numbers, similar to that of the drag-reducing flow. In this flow case, the results are compared with DNS data obtained by Coleman et al. [43] (see **Table 2**), which uses a Fourier-Legendre spectral discretization along with a hybrid implicit-explicit third-order four-substep time-advance algorithm. **Figure 2** compares the mean velocity and the Reynolds stresses distributions in wall units with the results in database by Coleman et al. [43]. It is seen that good agreements have been obtained for the mean velocity, in spite of the fact that the Reynolds stresses distributions are not accurately modeled. However, both the trends in wall-normal direction approximately coincide with the standard results, which provide convincing evidence for the effectiveness of the current DNS solver for compressible turbulent flows. Therefore, the spanwise oscillation velocity

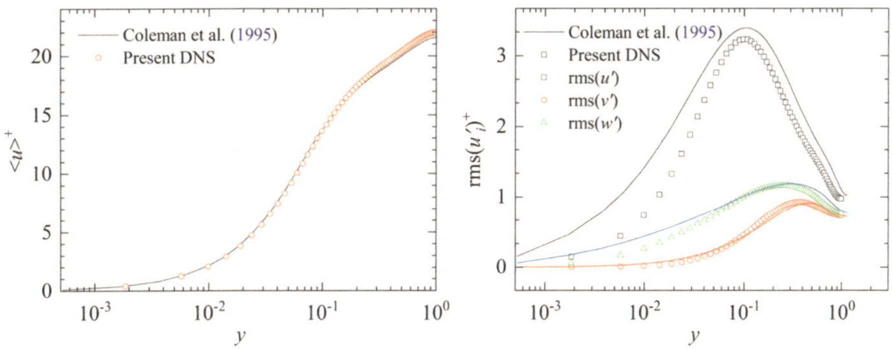

Figure 2.
Comparison of (a) mean streamwise velocity profile and (b) root mean square (rms) values of the velocity fluctuations in compressible turbulent channel flow for $Re_b = 3000$ *at* $Ma_b = 1.5$.

can be imposed on the walls to investigate the drag reduction performance for compressible turbulent channel flows.

2.2 Assessment of turbulent drag reduction via SWO

For compressible turbulent channel flows, we have established different oscillatory cases to investigate the effects of maximal wall velocity $W_m{}^+$ and oscillation period T^+, as shown in **Table 3**. The friction coefficient C_f is the ratio between the boundary mean shear stress and the mean kinetic energy per unit volume. With nondimensionalized quantities, this can be written as,

$$C_f = \frac{2}{\mathrm{Re}} \frac{\partial (u)}{\partial y}\bigg|_{y=1,1}, \tag{7}$$

where $Re = 2h\rho_0 U_b/\mu(T_0)$ is the bulk Reynolds number, ρ_0 is the initial uniform density, and $\mu(T_0)$ is the molecular viscosity at the wall temperature T_0.

Based on the work by Gomez et al. [53], for a compressible turbulent flow, we need to define two averaging operators. The Favre average permits us to transform the mean of a product into the product of a mean as noted by Coleman et al. [43]. In the case of the channel flow, the Reynolds operator consists in averaging on x-z space

Cases	$W_m{}^+$	T^+	10^3C_f	$\%P_{\mathrm{sav}}$	10^3C_L	10^3C_T	10^3C_C	10^3C_{CT}
cCase0	0	0	7.81 (7.7)	0	1.87	5.61	0.33	0.00103
cW6T100	6	100	5.99	23.3	1.870	3.94	0.24	0.00966
cW12T50	12	50	5.59	28.4	1.78	3.55	0.25	0.0112
cW12T100	12	100	5.42	30.6	1.71	3.57	0.12	0.0187
cW12T200	12	200	5.92	24.1	1.76	3.98	0.16	0.0205
cW18T100	18	100	1.72	39.3	1.62	3.02	0.071	0.0261

Table 3.
Power budget data for compressible oscillatory turbulent channel flows for $Re_b = 3000$. *The data inside the braces are the results from Modesti and Pirozzoli [52].*

variables and t time variable. The single prime and the double prime denote the turbulent fluctuations with respect to Reynolds and Favre averages, respectively. The difference between the Reynolds-averaged and the Favre-averaged quantities can be written as,

$$\langle f\rangle - \{f\} = \left\langle f^{''}\right\rangle = -\frac{\langle \rho' f'\rangle}{\langle \rho\rangle} = -\frac{\left\langle \rho' f''\right\rangle}{\langle \rho\rangle}. \tag{8}$$

The difference between Reynolds-averaged and Favre-averaged quantities is mainly relevant in the near-wall region by using DNS [50]. Therefore, the operator definitions give the following relations:

$$\frac{\partial\langle\cdot\rangle}{\partial x} = \frac{\partial\langle\cdot\rangle}{\partial z} = \frac{\partial\langle\cdot\rangle}{\partial t} = 0, \tag{9}$$

$$\frac{\partial\{\cdot\}}{\partial x} = \frac{\partial\{\cdot\}}{\partial y} = \frac{\partial\{\cdot\}}{\partial t} = 0. \tag{10}$$

Hereafter, by assuming (i) constant flow rate, (ii) homogeneity in the streamwise (x) and the spanwise (z) directions, (iii) no-slip conditions at the wall surfaces $y = -1$ and $+1$, and (iv) symmetry with respect to the center plane. Then, the momentum equation averaged along homogeneous directions x-z and time t variables yields,

$$\frac{\partial <\rho> \{uw\}}{\partial y} = \frac{1}{\mathrm{Re}}\frac{\partial <\tau_{xy}>}{\partial y} - f_1, \tag{11}$$

where f_1 is the force gradient term, defined by the following relation:

$$f_1 = -\frac{1}{2}\left(\frac{2}{\mathrm{Re}}\frac{\partial\langle u\rangle}{\partial y}\bigg|_{y=-1,1}\right) = -\frac{1}{2}C_f \tag{12}$$

We can finally obtain the following relationship:

$$\begin{aligned} C_f = \underbrace{\frac{6}{\mathrm{Re}}}_{C_L} + \underbrace{6\int_{-1}^{0} y\langle\rho\rangle\{u''v''\}dy}_{C_T} + \underbrace{\frac{6}{\mathrm{Re}}\int_{-1}^{0} -y\langle\tilde{\mu}\rangle\frac{\partial\langle u\rangle}{\partial y}dy}_{C_C} \\ + \underbrace{\frac{6}{\mathrm{Re}}\int_{-1}^{0} -y\left\langle \mu'\left(\frac{\partial u'}{\partial y} + \frac{\partial w'}{\partial y}\right)\right\rangle dy.}_{C_{CT}} \end{aligned} \tag{13}$$

This relation shows that the skin friction coefficient can be split into four contributing terms: the laminar contribution C_L, the turbulent contribution C_T, the compressible contribution C_{CT}, and the compressible-turbulent interaction term C_{CT}. The turbulent term C_T is proportional to the weighted average of the Reynolds stress, where the weight is linearly decreasing with the distance from the wall. As remarked by Fukagata et al. [54], this explains why the frictional drag observed in wall turbulence is mainly due to the turbulence wall structures which occur closer to the wall than the position of the maximum Reynolds stress. The compressible term C_C is proportional to the mean viscosity fluctuation due to thermal variations and the mean wall normal velocity gradient. This term is obtained by a weighted average with the linearly decreasing weight as for C_T. The last contributing term C_{CT} is proportional to the weighted average of the mean product between viscosity and velocity fluctuation gradients. The skin friction coefficient with the four components for different cases are given in **Table 3**.

It is clear that the measured friction coefficient for nonoscillatory case (cCase0) compared very well with the value reported by Modesti and Pirozzoli [52] (within a 0.2% difference). This result also agrees well with the simulation data by Gomez et al. [53] (within a 1% difference). In addition, **Figure 3** presents $P_{sav}\%$ as a function of maximum wall velocity ${W_m}^+$ and oscillation period T^+. The total drag-reduction performance as a function of ${W_m}^+$ and T^+, is similar to that in incompressible flows, namely for a given value of ${W_m}^+$, the highest drag reductions are attained at $T^+ = 100$, and that drag reduction appears to increase monotonically with ${W_m}^+$ for a fixed period of oscillation. These findings are in good agreement with the observations by Fang et al. [37, 38] and Ni et al. [39].

Figure 4 further shows the four components of the skin friction coefficient for a function of maximum wall velocity ${W_m}^+$ and oscillation period T^+. It is clear that there is a remarkable decrease of the turbulent contribution C_T under the action of oscillatory walls, and this tendency is the most obvious for the cases with larger ${W_m}^+$ at $T^+ = 100$.

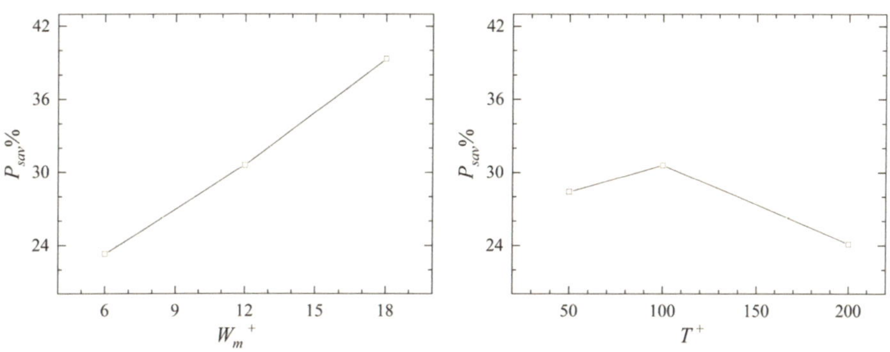

Figure 3.
$\%P_{sav}$ as a function of (a) maximum wall velocity ${W_m}^+$ at $T^+ = 100$ and of (b) oscillation period T^+ at ${W_m}^+ = 12$.

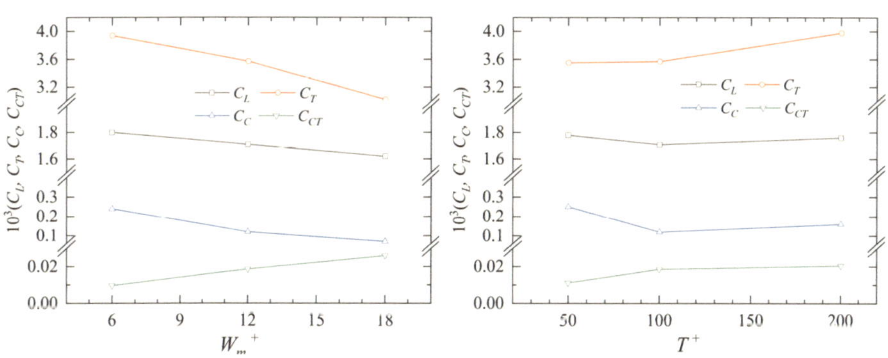

Figure 4.
The four components of skin friction coefficient C_L, C_T, C_C, C_{CT} as a function of (a) maximum wall velocity ${W_m}^+$ at $T^+ = 100$ and of (b) oscillation period T^+ at ${W_m}^+ = 12$.

The turbulent contribution dominates the overall drag reduction performance, in spite of the fact that the compressible term C_C due to thermal variations and the mean wall normal velocity gradient slightly decreases with the increasing ${W_m}^+$. This implies that the compressible effect is modest with respect to the turbulent effect for the skin-fiction in compressible turbulent channel flows, which is identical to the findings by Gomez et al. [53]. In addition, this phenomenon also repeats for the laminar contribution C_L. Interestingly, the compressible-turbulent interaction term C_{CT} slightly increases for larger ${W_m}^+$. As remarked by Fukagata et al. [54] and Li et al. [55], the frictional drag observed in wall turbulence is mainly due to the turbulence wall structures which occur closer to the wall than the position of the maximum Reynolds stress. The compressible term is proportional to the mean viscosity fluctuation due to thermal variations and the mean wall normal velocity gradient [56]. This term is obtained by weighted average with the linearly decreasing weight as for the turbulent term. In addition, the compressible contribution to the skin friction is concentrated in a region closer to the wall than that of the Reynolds stress contribution, which is more largely extended through the channel. This can be physically explained by the fact that the thermal boundary layer is thinner than the kinetic boundary layer.

3. Challenges encountered

DNS is a simulation in computational fluid dynamics in which the Navier–Stokes equations are numerically solved without any turbulence model. This means that the whole range of spatial and temporal scales of the turbulence must be resolved. All the spatial scales of the turbulence must be resolved in the computational grid, from the smallest dissipative scales (Kolmogorov microscales), up to the integral scale, associated with the motions containing most of the kinetic energy. Therefore, the computational cost of DNS is very high, even at low Reynolds numbers. For the cases in our simulations, we have more than 10 million grid points. The simulations for compressible cases were run for more than 5.0×10^5 CPU core hours to get statistically stable results.

The methodology for solving incompressible flow problems using pressure-based algorithms has been well developed. For our studies, these algorithms have to be extended to allow for the simulation of compressible flows in various Mach number regimes. Incompressible flow solutions do not generally require the solution of the energy equation. However, compressibility affects both hydrodynamics and thermodynamics, necessitating the simultaneous solution of the continuity, momentum, and energy equations.

The dependence of density on pressure and temperature, which is expressed via an equation of state, further complicates the velocity-pressure coupling present in incompressible flows. The derivation of the pressure correction equation now involves a density correction that introduces a convection-like term to the equation, in addition to the diffusion-like term introduced by the velocity correction. Another difficulty is the complex boundary conditions introduced that arise in compressible flow problems.

4. Conclusions and recommendations

We have established a reliable DNS model for compressible turbulent channel flow, and the skin-friction drag reduction by the spanwise motion generated by the wall movement has been studied. The aim is to bring skin-friction drag reduction control by spanwise wall oscillation into real engineering applications. It is found that the reduced drag reduction (DR) is mainly due to the decrease turbulent contribution under the action of oscillatory walls. Additionally, the small compressible term, related to thermal variations and the mean wall normal velocity gradient, also slightly decreases with increasing ${W_m}^+$.

However, more DNS cases with different maximum wall velocity, oscillation period, and flow Reynold number for compressible cases should be analyzed. The optimal combination with highest drag reduction efficient has a significant importance on real applications, which deserves to be studied in detail to characterize the underlying mechanisms. In addition, the near-wall streaks are conditioned from the turbulent fields. Understanding the modulation of the exact coherent structure by the spanwise wall oscillation can provide a direct view of how the regeneration cycle is weakened by the spanwise motion. At the flight Reynolds number, the contribution to skin-friction from the very large scale motions is very important, and it is an interesting topic to understand how to control the very large scale motions effectively from the outer region for compressible flows.

Acknowledgements

The authors gratefully acknowledge the financial support from the QCY Innovative and Entrepreneurial Talent Programme of Shaanxi Province and the Young Talent Support Plan of Xi'an Jiaotong University - XJTU. Computational resources were granted by the High-Performance-Computer Center of XJTU.

Conflict of interest

The authors declared that there is no conflict of interest.

Author details

Wenjun Yuan*†, Dongming Chen and Haifeng Zhang
School of Chemical Engineering and Technology, Xi'an Jiaotong University, Xi'an, China

*Address all correspondence to: wenjun.yuan@xjtu.edu.cn

† These authors contributed equally.

References

[1] Kim J. Physics and control of wall turbulence for drag reduction. Philosophical Transactions of the Royal Society A: Mathematical, Physical and Engineering Sciences. 2011;**369**: 1396-1411. DOI: 10.1098/rsta.2010.0360

[2] Choi H, Moin P, Kim J. Active turbulence control for drag reduction in wall-bounded flows. Journal of Fluid Mechanics. 1994;**262**:75-110

[3] Sirovich L, Karlsson S. Turbulent drag reduction by passive mechanisms. Nature. 1997;**388**:753-755. DOI: 10.1038/41966

[4] White CM, Mungal MG. Mechanics and prediction of turbulent drag reduction with polymer additives. Annual Review of Fluid Mechanics. 2008;**40**:235-256. DOI: 10.1146/annurev.fluid.40.111406.102156

[5] Checco A, Ocko BM, Rahman A, Black CT, Tasinkevych M, Giacomello A, et al. Collapse and reversibility of the superhydrophobic state on nanotextured surfaces. Physical Review Letters. 2014; **112**:216101. DOI: 10.1103/PhysRevLett.112.216101

[6] Tay C, Khoo B, Chew Y. Mechanics of drag reduction by shallow dimples in channel flow. Physics of Fluids. 2015;**27**: 035109. DOI: 10.1063/1.4915069

[7] Rastegari A, Akhavan R. The common mechanism of turbulent skin-friction drag reduction with superhydrophobic longitudinal microgrooves and riblets. Journal of Fluid Mechanics. 2018;**838**: 68-104. DOI: 10.1017/jfm.2017.865

[8] Jung W-J, Mangiavacchi N, Akhavan R. Suppression of turbulence in wall-bounded flows by high-frequency spanwise oscillations, physics of fluids a. Fluid Dynamics. 1992;**4**:1605-1607. DOI: 10.1063/1.858381

[9] Akhavan R, Jung W, Mangiavacchi N. Turbulence control in wall-bounded flows by spanwise oscillations. In: Advances in Turbulence IV: Proceedings of the Fourth European Turbulence Conference 30th June–3rd July 1992. Dordrecht: Springer; 1993. pp. 299-303. DOI: 10.1007/978-94-011-1689-3_48

[10] Kim J, Moin P, Moser R. Turbulence statistics in fully developed channel flow at low Reynolds number. Journal of Fluid Mechanics. 1987;**177**:133-166. DOI: 10.1017/S0022112087000892

[11] Baron A, Quadrio M. Turbulent drag reduction by spanwise wall oscillations. Applied Scientific Research. 1995;**55**: 311-326. DOI: 10.1007/BF00856638

[12] Laadhari F, Skandaji L, Morel R. Turbulence reduction in a boundary layer by a local spanwise oscillating surface. Physics of Fluids. 1994;**6**: 3218-3220. DOI: 10.1063/1.868052

[13] Di Cicca GM, Iuso G, Spazzini PG, Onorato M. Particle image velocimetry investigation of a turbulent boundary layer manipulated by spanwise wall oscillations. Journal of Fluid Mechanics. 2002;**467**:41-56. DOI: 10.1017/S002211200200157X

[14] Ricco P. Modification of near-wall turbulence due to spanwise wall oscillations. Journal of Turbulence. 2004;**5**:024. DOI: 10.1088/1468-5248/5/1/024

[15] Choi K-S, Graham M. Drag reduction of turbulent pipe flows by circular-wall oscillation. Physics of Fluids. 1998;**10**: 7-9. DOI: 10.1063/1.869538

[16] Choi J-I, Xu C-X, Sung HJ. Drag reduction by spanwise wall oscillation in wall-bounded turbulent flows. AIAA Journal. 2002;**40**:842-850. DOI: 10.2514/2.1750

[17] Quadrio M, Ricco P. Critical assessment of turbulent drag reduction through spanwise wall oscillations. Journal of Fluid Mechanics. 2004;**521**: 251-271. DOI: 10.1017/S0022112004001855

[18] Moarref R, Jovanović MR. Model-based design of transverse wall oscillations for turbulent drag reduction. Journal of Fluid Mechanics. 2012;**707**: 205-240. DOI: 10.1017/jfm.2012.272

[19] Ricco P, Quadrio M. Wall-oscillation conditions for drag reduction in turbulent channel flow. International Journal of Heat and Fluid Flow. 2008;**29**: 891-902. DOI: 10.1016/j.ijheatfluidflow.2007.12.005

[20] Xu C-X, Huang W-X. Transient response of Reynolds stress transport to spanwise wall oscillation in a turbulent channel flow. Physics of Fluids. 2005;**17**: 018101-018101. DOI: 10.1063/1.1827274

[21] Duggleby A, Ball KS, Paul MR. The effect of spanwise wall oscillation on turbulent pipe flow structures resulting in drag reduction. Physics of Fluids. 2007;**19**:125107. DOI: 10.1063/1.2825428

[22] Ricco P, Ottonelli C, Hasegawa Y, Quadrio M. Changes in turbulent dissipation in a channel flow with oscillating walls. Journal of Fluid Mechanics. 2012;**700**:77-104. DOI: 10.1017/jfm.2012.97

[23] Touber E, Leschziner MA. Near-wall streak modification by spanwise oscillatory wall motion and drag-reduction mechanisms. Journal of Fluid Mechanics. 2012;**693**:150-200. DOI: 10.1017/jfm.2011.507

[24] Agostini L, Touber E, Leschziner MA. Spanwise oscillatory wall motion in channel flow: Drag-reduction mechanisms inferred from dns-predicted phase-wise property variations at. Journal of Fluid Mechanics. 2014;**743**:606-635. DOI: 10.1017/jfm.2014.40

[25] Yakeno A, Hasegawa Y, Kasagi N. Modification of quasi-streamwise vortical structure in a drag-reduced turbulent channel flow with spanwise wall oscillation. Physics of Fluids. 2014; **26**:085109. DOI: 10.1063/1.4893903

[26] Gatti D, Quadrio M. Reynolds-number dependence of turbulent skin-friction drag reduction induced by spanwise forcing. Journal of Fluid Mechanics. 2016;**802**:553-582. DOI: 10.1017/jfm.2016.485

[27] Dhanak M, Si C. On reduction of turbulent wall friction through spanwise wall oscillations. Journal of Fluid Mechanics. 1999;**383**:175-195. DOI: 10.1017/S0022112098003784

[28] Galionis I, Hall P. On the stabilization of the most amplified görtler vortex on a concave surface by spanwise oscillations. Journal of Fluid Mechanics. 2005;**527**:265-283. DOI: 10.1017/S0022112004003118

[29] Negi PS, Mishra M, Skote M. Dns of a single low-speed streak subject to spanwise wall oscillations. Flow, Turbulence and Combustion. 2015; **94**:795-816. DOI: 10.1007/s10494-015-9599-z

[30] Blesbois O, Chernyshenko SI, Touber E, Leschziner MA. Pattern prediction by linear analysis of turbulent flow with drag reduction by wall

oscillation. Journal of Fluid Mechanics. 2013;**724**:607-641. DOI: 10.1017/jfm.2013.165

[31] Agostini L, Leschziner M. The impact of footprints of large-scale outer structures on the near-wall layer in the presence of drag-reducing spanwise wall motion. Flow, Turbulence and Combustion. 2018;**100**:1037-1061. DOI: 10.1007/s10494-018-9917-3

[32] Karniadakis G, Choi K-S. Mechanisms on transverse motions in turbulent wall flows. Annual Review of Fluid Mechanics. 2003;**35**:45-62. DOI: 10.1146/annurev.fluid.35.101101.161213

[33] Lagha M, Kim J, Eldredge JD, Zhong X. Near-wall dynamics of compressible boundary layers. Physics of Fluids. 2011;**23**:065109

[34] Hamilton JM, Kim J, Waleffe F. Regeneration mechanisms of near-wall turbulence structures. Journal of Fluid Mechanics. 1995;**287**:317-348

[35] Schoppa W, Hussain F. Coherent structure generation in near-wall turbulence. Journal of Fluid Mechanics. 2002;**453**:57-108

[36] Jiménez J, Pinelli A. The autonomous cycle of near-wall turbulence. Journal of Fluid Mechanics. 1999;**389**:335-359

[37] Fang J, Lu L, Shao L. Large eddy simulation of compressible turbulent channel flow with spanwise wall oscillation. Science in China Series G: Physics, Mechanics and Astronomy. 2009;**52**:1233-1243. DOI: 10.1007/s11433-009-0165-3

[38] Fang J, Lu L. Large eddy simulation of compressible turbulent channel flow with active spanwise wall fluctuations. Modern Physics Letters B. 2010;**24**:1457-1460. DOI: 10.1142/S0217984910023864

[39] Ni W, Lu L, Ribault CL, Fang J. Direct numerical simulation of supersonic turbulent boundary layer with spanwise wall oscillation. Energies. 2016;**9**:154. DOI: 10.3390/en9030154

[40] Yao J, Hussain F. Toward vortex identification based on local pressure-minimum criterion in compressible and variable density flows. Journal of Fluid Mechanics. 2018;**850**:5-17. DOI: 10.1017/jfm.2018.465

[41] Yao J, Hussain F. Supersonic turbulent boundary layer drag control using spanwise wall oscillation. Journal of Fluid Mechanics. 2019;**880**:388-429

[42] Yao J, Hussain F. Mach number effect on drag control via spanwise wall oscillation in wall-bounded turbulent flows. Bulletin of the American Physical Society. 2018;**63**:13

[43] Coleman GN, Kim J, Moser RD. A numerical study of turbulent supersonic isothermal-wall channel flow. Journal of Fluid Mechanics. 1995;**305**:159-183. DOI: 10.1017/S0022112095004587

[44] Morinishi Y, Tamano S, Nakabayashi K. Direct numerical simulation of compressible turbulent channel flow between adiabatic and isothermal walls. Journal of Fluid Mechanics. 2004;**502**:273-308. DOI: 10.1017/S0022112003007705

[45] Foysi H, Sarkar S, Friedrich R. Compressibility effects and turbulence scalings in supersonic channel flow. Journal of Fluid Mechanics. 2004;**509**:207-216. DOI: 10.1017/S0022112004009371

[46] Sun Z, Ren Y, Larricq C. Drag reduction of compressible wall

turbulence with active dimples. Science China Physics, Mechanics and Astronomy. 2011;**54**:329-337. DOI: 10.1007/s11433-010-4189-5

[47] Modesti D, Pirozzoli S. Reynolds and mach number effects in compressible turbulent channel flow. International Journal of Heat and Fluid Flow. 2016;**59**: 33-49. DOI: 10.1016/j.ijheatfluidflow.2016.01.007

[48] Hoyas S, Jiménez J. Scaling of the velocity fluctuations in turbulent channels up to re τ= 2003. Physics of Fluids. 2006;**18**:011702. DOI: 10.1063/1.2162185

[49] Bernardini M, Pirozzoli S, Orlandi P. Velocity statistics in turbulent channel flow up to. Journal of Fluid Mechanics. 2014;**742**:171-191. DOI: 10.1017/jfm.2013.674

[50] Huang P, Coleman G, Bradshaw P. Compressible turbulent channel flows: Dns results and modelling. Journal of Fluid Mechanics. 1995;**305**:185-218. DOI: 10.1017/S0022112095004599

[51] Bernardini M, Pirozzoli S, Quadrio M, Orlandi P. Turbulent channel flow simulations in convecting reference frames. Journal of Computational Physics. 2013;**232**:1-6. DOI: 10.1016/j.jcp.2012.08.006

[52] Modesti D, Pirozzoli S. A low-dissipative solver for turbulent compressible flows on unstructured meshes, with openfoam implementation. Computers & Fluids. 2017;**152**:14-23. DOI: 10.1016/j.compfluid.2017.04.012

[53] Gomez T, Flutet V, Sagaut P. Contribution of Reynolds stress distribution to the skin friction in compressible turbulent channel flows. Physical Review E. 2009;**79**:035301. DOI: 10.1103/PhysRevE.79.035301

[54] Fukagata K, Iwamoto K, Kasagi N. Contribution of Reynolds stress distribution to the skin friction in wall-bounded flows. Physics of Fluids. 2002; **14**:L73-L76

[55] Li W, Fan Y, Modesti D, Cheng C. Decomposition of the mean skin-friction drag in compressible turbulent channel flows. Journal of Fluid Mechanics. 2019; **875**:101-123

[56] Wenzel C, Gibis T, Kloker M. About the influences of compressibility, heat transfer and pressure gradients in compressible turbulent boundary layers. Journal of Fluid Mechanics. 2022;**930**:A1

Chapter 4

Active Flow Control and Its Applications in Supersonic Boundary Layer

Gan Tian

Abstract

Boundary layer transition is a typical aerodynamic phenomenon in supersonic flow that refers to the transition of the boundary layer from laminar flow to turbulent flow. The instabilities and transition mechanisms are complicated. A turbulent boundary layer has an associated higher energy content more momentum and, therefore, is less prone to be separated. Furthermore, accurate and fast transition in the flow over the compressor blades in turbo machinery provides significant flow mixing, resulting in an increase in engine performance. In addition, the development of the scramjet made it possible to use plasma actuators as turbulator to increase the turbulence within an isolator and increase combustion efficiency. Thus, different methods of boundary layer transition have been proposed. The focus of this chapter is a low Reynolds number supersonic boundary layer and its interaction with an active flow control device. The Plasma-Based Actuator's high repetition rate enables itself as an unsteady control mechanism to address the unsteady flow features. Moreover, its strong forcing makes it suitable for flow control applications in the high-speed flow regime, especially the supersonic boundary layer.

Keywords: supersonic boundary layer, surface arc, laminar to turbulent transition, shock wave boundary layer interaction, active flow control

1. Introduction

Boundary layer transition is a typical aerodynamic phenomenon in supersonic flow that refers to the transition of the boundary layer from laminar flow to turbulent flow. The instabilities and transition mechanisms are summarized in detail in Ref. [1]. The effective control of the laminar-turbulent transition is crucial to increase the resistance to the adverse pressure gradient resulting from the shock in the shock wave/boundary layer interaction (SWBLI) in Ref. [2]. The major difference between laminar and turbulent flow is the velocity profile. A turbulent boundary layer has more momentum and is less prone to be separated by the disturbance. Furthermore, accurate and fast transition in the flow over the compressor blades in turbo machinery provides significant flow mixing, resulting in an increase in engine performance. In addition, the development of the scramjet made it possible to use plasma actuators as turbulator to increase the turbulence within an isolator and increase combustion efficiency. Thus, different methods of boundary layer transition have been proposed. In general, boundary layer transition can be classified into three categories based on different mechanisms that act separately or in combination: natural transition, bypass transition, and separation-induced transition. The transition path is summarized in Ref. [3], as shown in **Figure 1**. In either pathway, the final state is turbulence, which enhances mixing and leads to a higher-momentum boundary layer. The focus of this work is to investigate the ability of a plasma actuator array to produce disturbances in the transient growth for the purpose of promoting bypass transition.

Plasma-based actuators (PBAs) have attracted significant interest in the last three decades as an active flow control technology. Due to their simplicity, sufficient frequency bandwidth, fast response, and zero-drag compensation [4], they are superior to all passive control devices and several semiactive control devices, which are complex, heavy, noisy, and have vibrating moving parts. PBAs have various discharge

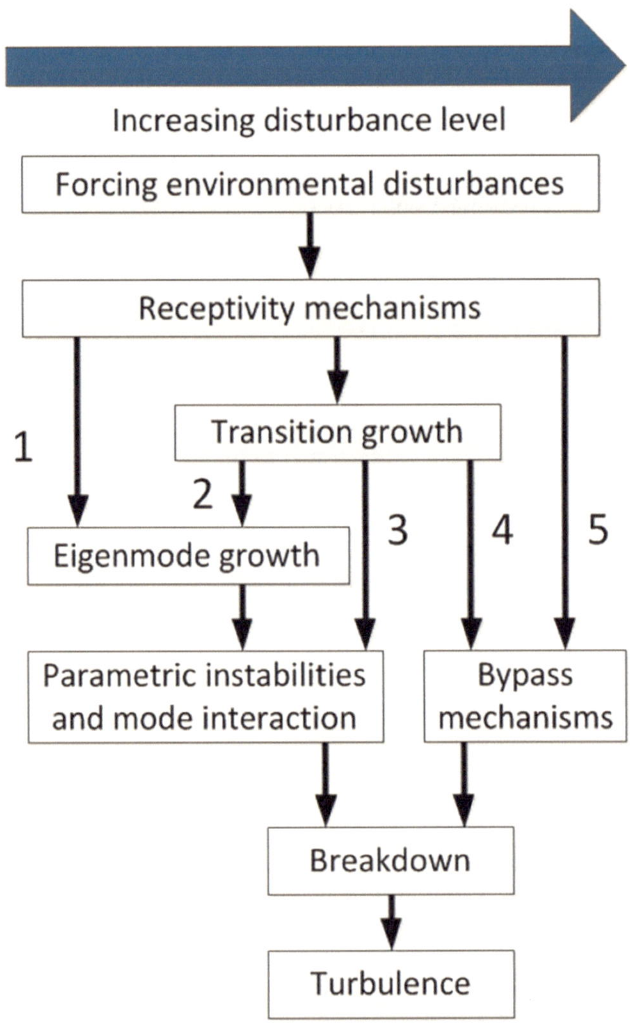

Figure 1.
Five boundary layer transition paths to turbulent flow.

types, such as a thermal perturbation induced by surface arc discharge in Ref. [5], wall-jet induced by dielectric barrier discharge in Ref. [6], and a high-speed synthetic jet induced by cavity discharge in Ref. [7], affecting the characteristics of the supersonic boundary layer. In particular, surface arc plasma actuators (SAPAs) are considered the most suitable for high-speed flow regimes due to their significant heating effect. Comprehensive reviews of these actuators can be found in Ref. [8–10].

Preliminary investigations of boundary layer control by PBAs have focused on transition delay in Ref. [11] because transition promotion is considered an undesirable effect in boundary layer control. Generally, a laminar boundary layer reduces the severity of heat transfer and drag friction. Therefore, a growing number of studies have focused on maintaining the laminar boundary layer state in Ref. [12, 13]. However, in some hyper- and supersonic applications, a long-duration laminar/laminar-turbulent transition region can cause detrimental impacts on vehicle performance. Thus, boundary layer transition promotion is also required. The basic method is to generate a fast turbulent spot with sufficient intensity and frequency. Experiment studies in Ref. [14, 15] have shown that both promotion and delay of boundary layer transition can be achieved by using a glow-discharge actuator. Two plasma actuators were used in a flat plate to influence the growth of the transition. The first actuator located downstream of the leading edge was operated in pulsed mode to create strong Tollmien-Schlichting (TS) waves, resulting in turbulence promotion. Two steadily operating actuators further downstream significantly damped the perturbations, suppressing the growth of the TS waves in the transition delay. However, the mainstream velocity was not specified.

In recent works [15, 16], SAPAs were used as a source of perturbations in supersonic flow, especially for a new circuit [17] with a parallel capacitance connection, which enables the PBAs to release sufficient intensity rapidly. This effect can lead to a rapid transition of the boundary layer to turbulent flow, either generating longitudinal streaks [18] or vortex structures that eventually develop into turbulent spots [19]. Recently, numerous publications have demonstrated the successful transition of the boundary layer to turbulent flow on a flat plate using plasma actuators based on surface arc discharge. The effect of steady and unsteady thermal perturbation on the transition of the boundary layer in a Mach 1.5 flat-plate flow is investigated in Ref. [20]. The results indicated that the transition process was enhanced by the actuators, suggesting that the perturbed boundary layer showed signs of transitioning to turbulence, although it remained in transitional flow at the end of the plate. More recently, the PBA's high repetition rate has been used as an unsteady control mechanism to address unsteady flow features. A SAPA array driven by altering the frequency was proposed in Ref. [21]. It can modify the boundary layer condition upstream by a ramp-induced separation shock in Mach 2 flow. Although the energy of a single pulse was very low, a significant vortex effect was observed in the boundary layer. After the activation of the plasma actuator, numerous periodic streamwise vortices were produced in the high-frequency actuation mode, resulting in the transition to turbulent flow of the boundary layer.

In this chapter, SAPA arrays are used in an experiment to perform boundary layer transition on a flat plate at Ma = 3. Several different frequency modes are examined. The ability to control the boundary layer transition is visualized by high-speed schlieren snapshots. Our objective is to demonstrate a new approach to using pulsed SAPAs to achieve fast transient growth in a supersonic laminar boundary layer flexibly and rapidly and reveal the underlying mechanism of supersonic boundary-layer transition by using this type of active flow control stratagem.

2. Supersonic boundary layer physics

The experimental data were collected in a supersonic wind tunnel with a fixed Ma number. The axisymmetric exit has a 300 mm outlet diameter and expands the flow to Mach 2.0 or 3.0 by using a different contoured nozzle. Two inspection windows are assembled on both sides of the test chamber for flow visualization. Therefore, optical measurement techniques and laser perturbations can have access to the core area of the test section. The wind tunnel is started by a circular membrane-breaking device with an electrical switch. The pump sets consist of three pump types, including water ring pumps (operating range from 0.06 bar to the local atmospheric pressure), sliding vane pumps (operating range from 0.03 to 0.08 bar), and root pumps (operating below 100 Pa). A schematic diagram of the wind tunnel structure is shown in **Figure 2** with a picture on the right.

The air inside the vacuum tank was pumped out to achieve a vacuum with a pressure of 5–10 Pa, enabling running times of 2.5 s. A stable supersonic flow with a Mach number of 3.0 was generated during the experiments, with a total pressure of 95.6 kPa and a total temperature of 296 K. In the core test region, the static pressure Ps and static temperature Ts were 2.61 kPa and 106 K, respectively. The velocity of the supersonic free stream was 618 m/s, resulting in a unit Reynolds number of $7.41 \times 106\ m^{-1}$. The operating parameters of this facility are listed in **Table 1**.

A Z-type light route was used in the schlieren investigations, as shown in **Figure 3**. A Gloria 500 W Xenon bulb provided illumination. The schlieren images were

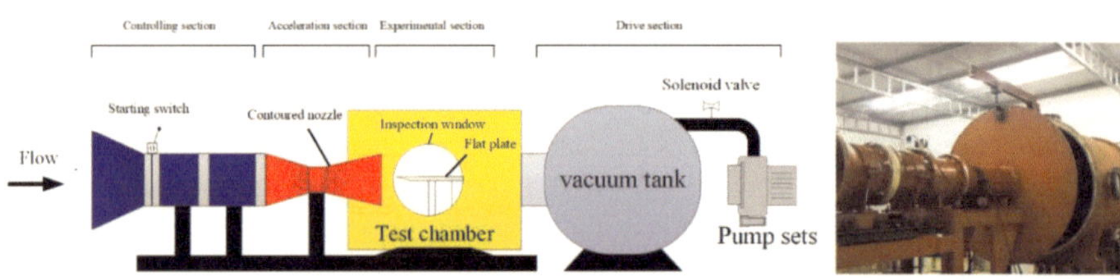

Figure 2.
Schematic diagram and picture of the supersonic wind tunnel structure.

Parameter	Description	Quantity
Ma	Free-stream Mach number	3
$P_0,(Pa)$	Stagnation pressure	95,600
$T_0(K)$	Stagnation temperature	296
$P_s(Pa)$	Static pressure	2612
$T_s(K)$	Static temperature	106
$\rho(kg/m^3)$	Density	0.0856
$U(m/s)$	Freestream velocity	618
$\mu(N{\cdot}s/m^2)$	Viscosity coefficient	7.43×10^{-6}
$Re(1/m)$	Reynolds number per meter	7.41×10^6
$t(s)$	Running time	2.5

Table 1.
Flow parameters.

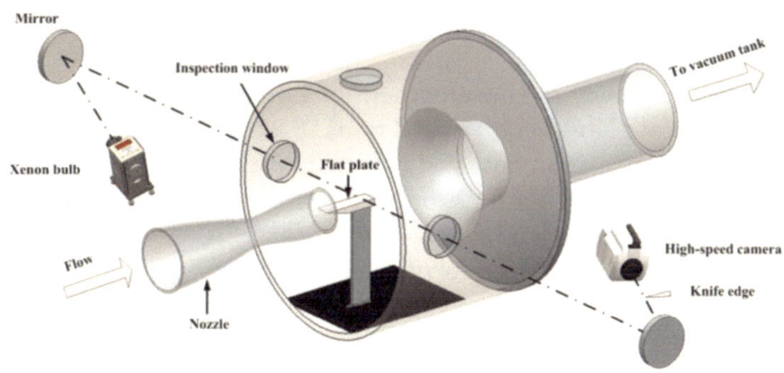

Figure 3.
Wind tunnel with diagnostic system.

captured with a Phantom V2512 ultra-high-speed camera with a maximum resolution of 1280 × 800 pixels. A field of view of 512 × 256 pixels was adopted in the experiment. The image acquisition frequency was 50 kHz with 1 μs minimum exposure time to freeze the boundary layer features. **Figure 4** shows a top view of the schlieren system.

Prior to the analysis, it is essential to investigate the morphology of a fully developed and undeveloped supersonic boundary layer and discuss the flow field. The boundary layer was developed on a flat plate with a sharp leading edge, installed in the

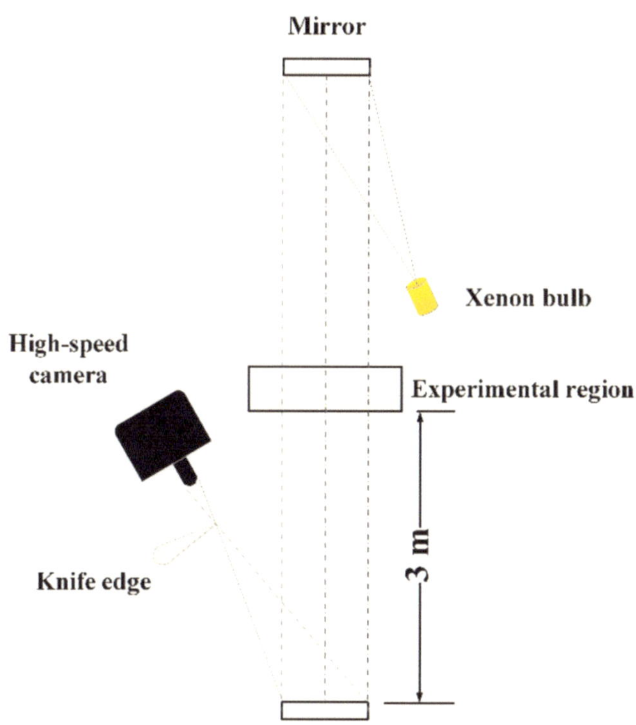

Figure 4.
Schlieren system (top view).

core test section at zero angle of attack. The dimensions of the plate were 440 × 110 × 20 mm.

Figure 5 shows images of the development of a natural boundary layer on a completely smooth flat plate (length = 440 mm) in a parallel flow at zero incidences without any intrusive measurements and plasma actuators. Note that the inspection window is smaller than the plate; thus, only part of the flow field can be captured. Therefore, the development of the supersonic boundary layer was displayed separately. **Figure 5(a)** shows the side view of the boundary layer ranging from x = 0 mm to x = 130 mm, corresponding to the leading edge of the flat plate (fore-plate). The boundary layer is very thin initially and appears almost uniform with no turbulent structure. It can be concluded that the supersonic boundary layer is in the laminar state. **Figure 5(b)** shows the side view ranging from x = 130 mm to x = 285 mm (mid-plate). The boundary layer appears as a straight, bright strip immediately above the flat surface; its growth instability leads to the formation of a small turbulent structure, corresponding to a laminar-turbulent transient boundary layer. Unlike in

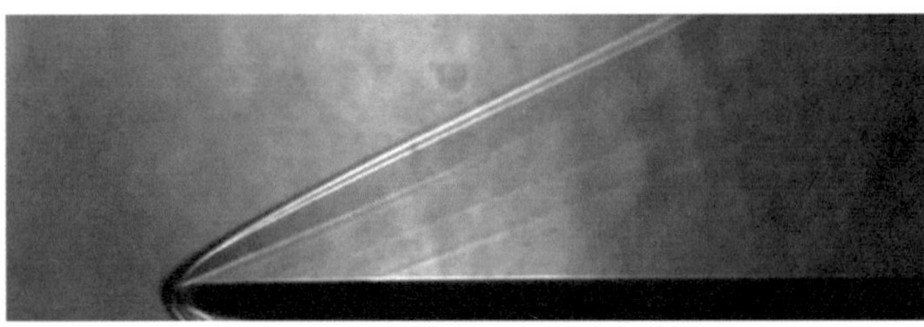

(a)

(b)

(c)

Figure 5.
Base flow on a flat plate (a) fore-plate (b) mid-plate (c) rear plate.

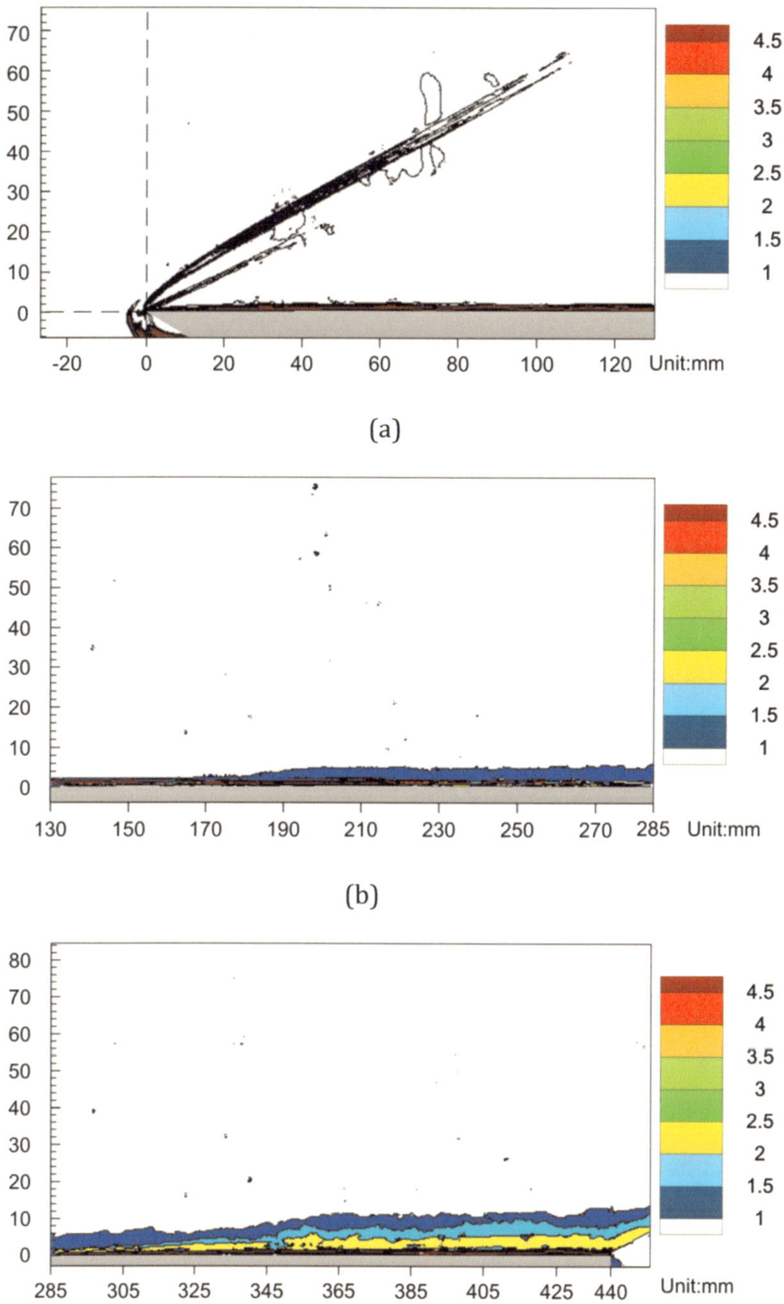

(c)

Figure 6.
Contours of the I_{rms} for boundary layer flow on a flat plate (a) I_{rms} intensity of the fore-plate (b) I_{rms} intensity of the mid-plate (c) I_{rms} intensity of the rear plate

Figure 5(a), turbulent flow is observed in **Figure 5(b)**. Thus, a natural transition occurs in the region x = 130 to 285 mm, after which the flow becomes fully turbulent. In the rear part of the flat plate (rear plate), ranging from x = 285 mm to x = 440 mm

(**Figure 1(c)**), a more rapid increase in the boundary layer instability occurs due to more large turbulent eddies in the near-wall region. Thus, a fully turbulent region is observed after x = 285 mm.

The I_{rms} for each part of the flat plate are shown in **Figure 6**. The leading edge of the flat plate is considered the origin of the coordinate system, with the positive x-axis pointing in the streamwise direction, and the positive y-axis pointing in the wall normal direction with respect to the surface. Due to the short camera exposure time and large image ensemble, the boundary layer is frozen with a shaped edge with respect to the mainstream flow, as shown in the contours of the I_{rms} (normalized magnitude) in **Figure 6**. The line connecting the lower I_{rms} value (above 1) along the wall can be taken as the border of the boundary layer. The contours of the I_{rms} for the fore plate are shown in **Figure 6(a)**, which agree with the fact in the raw schlieren image. The thickness of the boundary layer obtained from the I_{rms} has not increased, and its average value is very close to the theoretical value of a laminar boundary layer. The contours of I_{rms} also have the same magnitude in the wall-normal direction owing to the small adverse pressure gradient in the fore-plate. **Figure 6(b)** the I_{rms} contours have moved a considerable distance forward after 170 mm, and the width of I_{rms} from 1 to 1.5 count has thickened substantially. The raising width corresponds to the turbulent structures in the raw schlieren image, and their thickness is even larger downstream. Thus, the transition point can be determined as the point where the I_{rms} contours move up at approximately 170 mm. The final state of the boundary layer can be inferred from **Figure 6(c)**. Its thickness according to the I_{rms} continues to increase to a maximum value of 13.5 mm at the end of the flat plate. Two contours with higher I_{rms} values (ranging from 1.5 to 2 and from 2 to 2.5) move up after x = 305 mm, indicating significant vortical activity in the rear part of the flat plate. The I_{rms} is very high in the last region, indicating that a fully turbulent boundary layer has developed in the rear plate. The I_{rms} enables the analysis of the transitional region and the boundary layer state to quantify the effect of the SAPAs on boundary layer transition.

The supersonic boundary layer's response to a pressure gradient is also affected by viscous forces, which are often represented by the wall-shear stress or the skin-friction coefficient. The skin-friction coefficient for a flat-plate turbulent boundary layer is a significant function of the Mach number and decreases rapidly as the Reynolds number increases. Because shear forces tend to oppose the retardation effect, we can infer that the resistance of a laminar boundary layer to an imposed pressure gradient decreases with increasing Reynolds number. The situation is more subtle in the turbulent boundary layer, even at a moderate Mach number. Therefore, the behavior of boundary development is usually been characterized by assuming a perfect-fluid model.

3. Active flow control-surface arc plasma actuator

The surface arc plasma actuator has two main configurations, one is a horizontal arrangement, as shown in **Figure 7(a)**, where two cylindrical metal electrodes are flush mounted on an insulator to ensure that the surface is flat without disturbing the original boundary layer flow field. Another arrangement is a recessed layout, as shown in **Figure 7(b)**. The advantage of this layout is that the arc can be avoided blowing off by the incoming supersonic boundary layer flow. The disadvantage is that the cavity will have a certain disturbing effect on the original boundary layer flow field, which will easily confuse the effect of active flow control and cavity control. The two metal

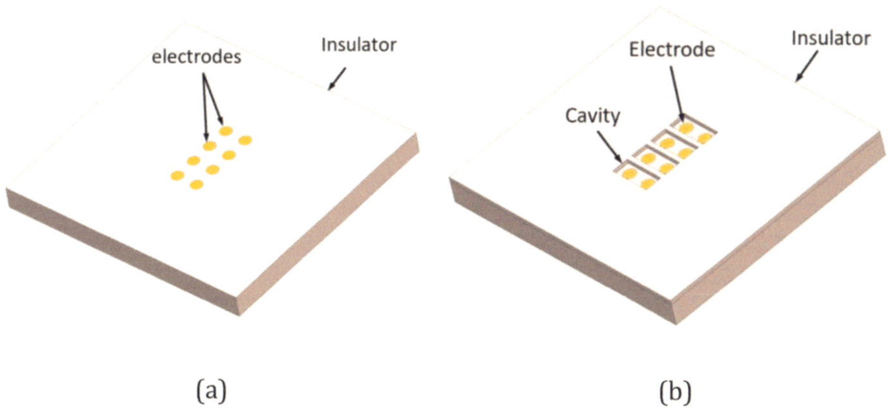

Figure 7.
Schematic diagram of the surface arc discharge actuator structure. (a) flush mounted layout (b) cavity layout.

electrodes are connected to the positive and negative side of the power supply, thus forming the cathode and anode, and through the load provided by the power supply, a potential difference is formed between the cathode and the anode, while an electromagnetic field is formed in the space above the surface. When the voltage applied between the electrodes exceeds a certain threshold, the air near the electrodes is pierced to form an arc-shaped discharge channel, emitting a dazzling white light and releasing electrical energy in the form of heat rapidly to form a transient explosion effect (as shown in **Figure 8**). Because of the arc-like shape of the discharge channel, this discharge is also known as an arc discharge. The discharge produces two types of disturbance structures, a shock wave and a localized thermal gas bubble. Because it is a high-voltage pulse discharge, there is a certain amount of electromagnetic interference (EMI) and anti-interference treatment is required. The most common method currently used is to wrap all the conductors in an electromagnetic protective wire jacket and install a grounding electrode near the discharge electrode to absorb most of the electrons. There is also a variant of the surface arc discharge actuator structure while the discharge electrode is placed in a concave cavity with a small hole open into the surface. As the application of the discharge, certain hot jets ejected from the hole to the bottom of the boundary layer as a source of disturbance. This active flow

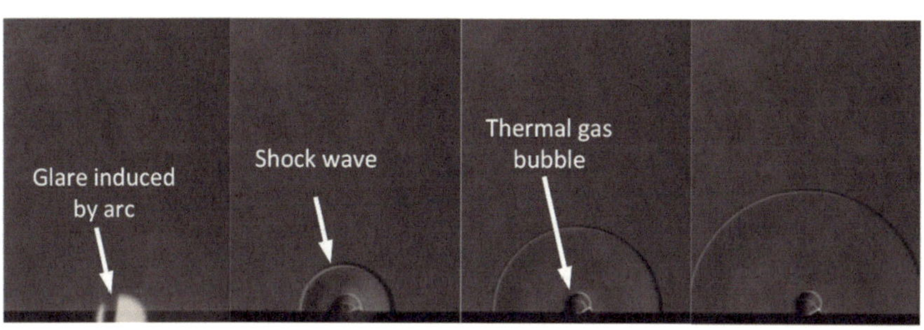

Figure 8.
The evolution of surface arc discharge in atmospheric air.

control device is also called plasma synthetic jet actuator, its effect on the control mechanism is also completely different from the surface arc plasma actuator.

A schematic diagram of the circuit system is shown in **Figure 9**. The external circuit system is powered by a high-voltage pulsed power supply and a voltage DC power supply. The high-voltage pulsed power supply is to produce a breakdown voltage to form a steady or pulsed arc between the anode and cathode. The voltage varies from a few kilovolts to tens of kilovolts depending on the ambient pressure. The DC power supply is to increase the energy deposition in the arc and to charge the capacitor beforehand.

Figure 10 shows the evolution of the arc pattern when using the above external circuit system with 1 mm electrode spacing. The DC power output voltage is 1.5 kV and the capacitance is 2 μF. To highlight the arc structure, the camera exposure time was reduced to obtain a black background, and the high-speed camera exposure time was set to 1.75 μs. t = 0 μs was set as the trigger of the actuation, and a bright white light was generated between the two electrodes in the figure (anode on the left and cathode on the right) indicating the generation of the surface arc. The maximum arc light intensity and arc area are reached at t = 2.49 μs after the discharge is generated. As the voltage decreases rapidly, the arc light fades and looks like a dumbbell shape, indicating that the arc energy is gradually dissipating into the air. The intensity of the

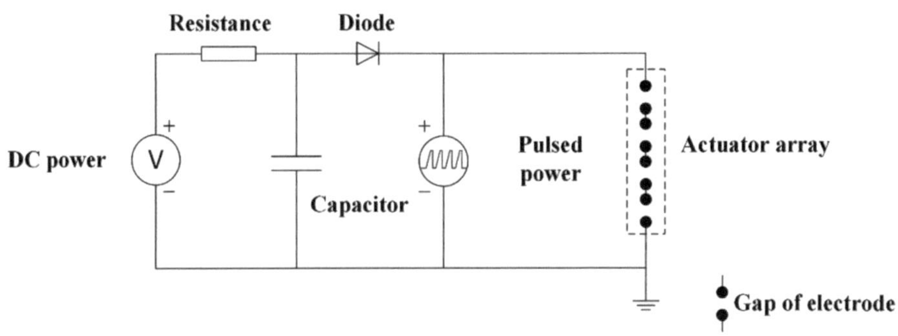

Figure 9.
External circuit system.

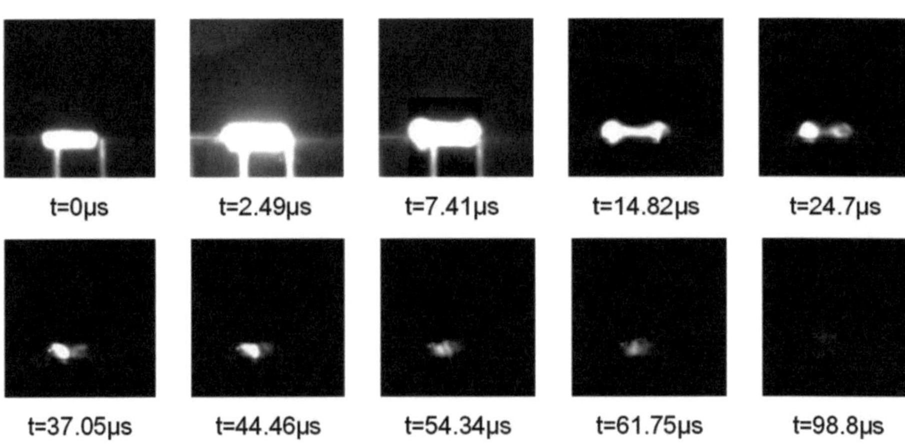

Figure 10.
The evolution of surface arc.

spot near the cathode decays more rapidly than that near the anode. At t = 98.8 μs, the bright light is not visible and the discharge process is considered to have been terminated.

Surface arc discharge is a rapid process of air breakdown to form a steady or pulsed arc between the anode and cathode. TGB and precursor shock (PS) are two major macroscopic characteristics that are generated in the discharging process while the former is mainly influenced by atmospheric pressure [22] and arc length. **Figure 11** shows the evolution of TGB and PS produced by a single SAPA with three different arc lengths. The measurements were performed at the same low pressure (12.24 kPa) as during the wind tunnel operation and the actuation frequency is 500 Hz. Here the variation in arc length is obtained by increasing the gap ΔZ between two electrodes. It should be noted that the maximum distance between electrodes was verified to be 10 mm at low ambient pressure, beyond which it can hardly break down the air to form an arc discharge. Thus the ΔZ between two electrodes was selected to be 3, 5, and 10 mm for case_1, case_2, and case_3, respectively. These images are captured with an exposure time of 1 μs and a frame size of 640 × 256 pixels, respectively. The resolution of the images is 0.93 mm/pixel, and the interval is about 10 μs. The recording is triggered at the start of the discharge using a DG 535 synchronous controller. The TGB is visible at 30 μs because, at prior times, it is overlapped by the arc light glow and characterized by a clean regular elliptical hemisphere compared to the following snapshots. About 260 μs after the trigger of discharge, more turbulent structures and sharper boundaries inside the TGB can be observed. Later, it becomes more turbulent in the last two snapshots and slowly grows upward.

As the discharge occurs in a Ma = 2 flow, a dazzling white light can be seen at first, as shown the **Figure 12**. The shock wave and the thermal gas bubble also can be seen in the schlieren image. At this moment, the internal structure of the thermal gas bubble is clear, with a small height and a regular oval shape, moving downstream with

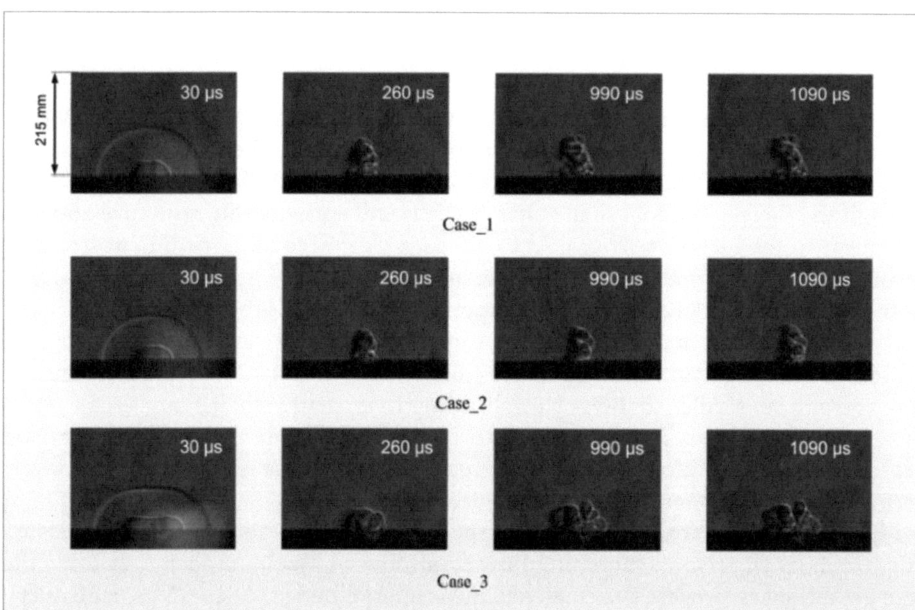

Figure 11.
Evolutions of TGB for different Δ Z in quiescent air.

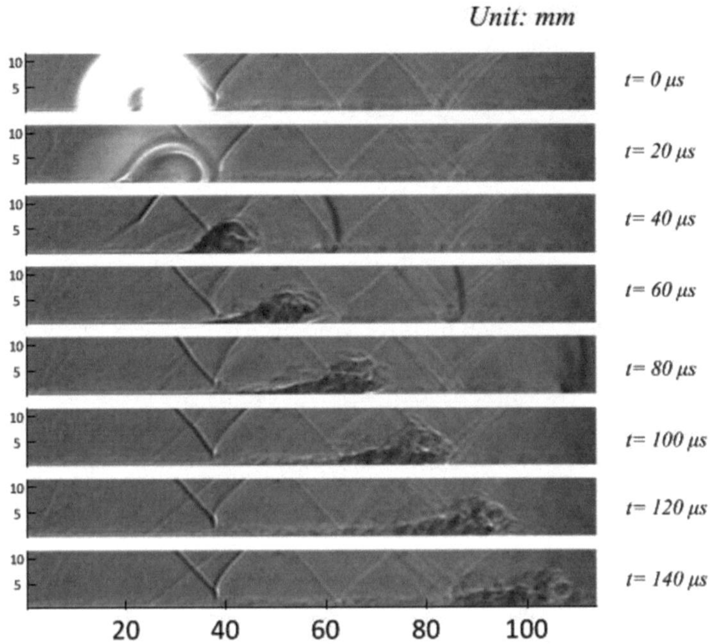

Figure 12.
Evolutions of TGB on a flat plate in a supersonic flow (ma = 2).

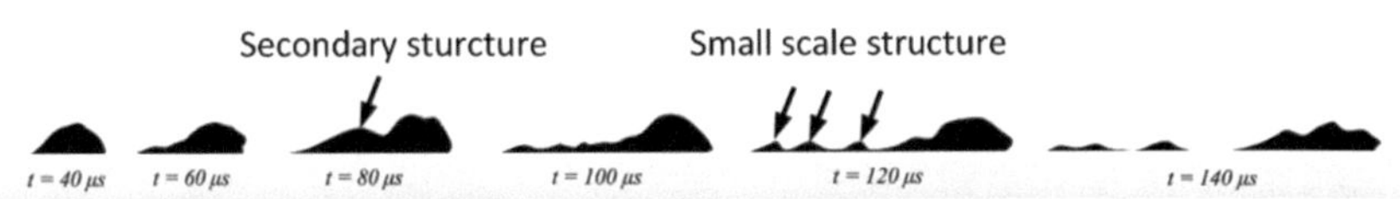

Figure 13.
Spatial gradient evolution of the TGB evolution.

the incoming flow with a propagation speed of about 520 m/s. The morphology of the thermal gas bubble transformed to turbulent because of the shear effect.

Figure 13 shows the diagram of the TGB evolution from 40 to 140 μs. At t = 40 μs, although the thermal bubble shape changes, it is still a monolithic structure and does not appear to shed larger vortices. At t = 80 μs, a clear secondary turbulent structure can be seen. The thermal bubble has little momentum in the generation process and the thickness is much greater than the supersonic boundary layer. The upper structure is subjected to higher shear stress and continues to move forward. The lower structure is subjected to shear from the upper layer and hysteresis occurs to form a secondary vortex structure, similar to the vortex shedding that occurs when a large-scale turbulent structure is subjected to a strong disturbance. In addition, the secondary structure breaks down further into smaller-scale structures due to free shear. Thus, the original thermal gas bubble is compressed in the vertical direction.

Figure 14 give typical PLS images of the thermal bubble for a surface arc plasma actuator under supersonic incoming flow conditions at Mach number 2, yielding the same evolutionary process that matches the schlieren image. At t = 25 μs, a distinct hemispherical black spot can be observed downstream of the actuator location, which can be determined as the hot bubble structure, as the hot bubble is hotter and denser, where the applied molecule particles rapidly condense to form a high density shadow,

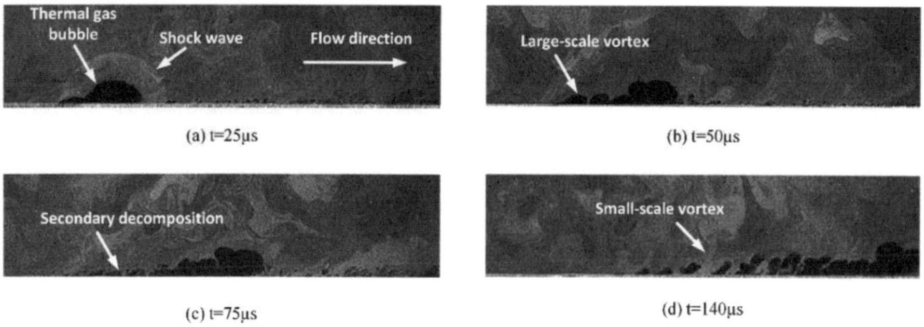

Figure 14.
PLS results of thermal gas bubbles under supersonic incoming flow conditions (Ma = 2).

which in turn appears different from the surrounding black area in the PLS image. The shock wave is also accurately captured by the PLS technique due to its high density and can be clearly seen in the image as an umbrella-shaped shock wave above the thermal gas bubble. At t = 25 μs the shape of the thermal bubble starts to change and a small number of large-scale vortex structures are derived, these also show the same black spots as the original thermal bubble.

4. Applications of surface arc plasma actuator on supersonic boundary layer

In this section, SAPA arrays are used in an experiment to perform boundary layer transition on a flat plate at Ma = 3. Three different frequency modes are examined. The ability to control the boundary layer transition is visualized by high-speed schlieren snapshots and semiquantitatively evaluated by image processing. Our objective is to demonstrate a new approach to using pulsed SAPAs to achieve transient growth in a laminar boundary layer flexibly and rapidly and reveal the underlying mechanism of boundary-layer transition.

The image of the first two pulses is selected to demonstrate the flow evolution of the laminar boundary layer with 10 kHz actuation frequency, as shown in **Figure 15**. The images of the flow field were acquired at 50,000 fps with an exposure time of 1 μs. The first snapshot captures the glare from the electric arc, which normally lasts a few microseconds, and the temporal delay between the first image and the high-voltage trigger should be a few microseconds. Thus, it should be noted that the first image at t = 0 does not overlap precisely with the high-voltage trigger of the SAPAs. Thus, the first snapshot is used as the temporal origin for the discussion. At t = 0 μs, tiny thermal gas bubbles (TGBs) have already been produced around the electrodes because of the local Joule heating effect, but the laminar boundary layer is unchanged. At t = 20 μs, the discharge process is complete, and the TGBs with a height of y = 0.2 mm have traveled around x = 90 mm downstream from the actuator. The precursor shockwave (PS) is expanding at Ma = 1.0, forming an umbrella shape above the TGBs. In the following snapshots, until t = 80 μs, the laminar nature of the boundary layer is unaffected because no distinct turbulent pattern is observed at the surface of the flat plate. As a result, it is assumed that the SAPAs did not affect the transition of the boundary layer within a certain distance downstream of the actuators. After propagating at about 100 μs, the flickering wakes of the TGBs indicate the generation of trailing vortices as the second pulse is observed in **Figure 15(f)**.

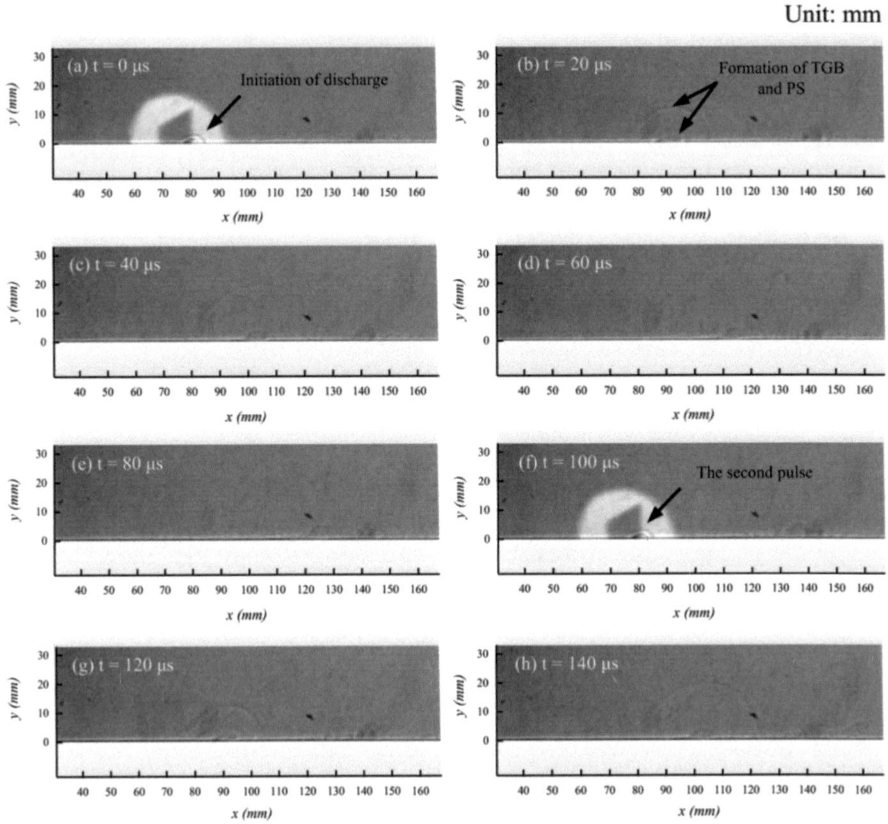

Figure 15.
Schlieren image of the boundary layer development with actuation in a Ma = 3 flow.

At t = 120 μs, the TGBs have traveled a distance of approximately 60 mm downstream of the actuators and have elongated into a long strip shape, with the major axis pointing toward the upper-right quarter. Trailing vortices are produced following the primary TGBs, contributing to a boundary layer with large displacement thicknesses and higher freestream streamlines. In the last snapshot, the TGBs appear as a turbulent structure, unlike in the other images. They move a distance of 80 mm and eventually leave the field of view.

The I_{rms} for 10 kHz reputation is shown in **Figure 16**. It has a range of 1 to 2.4 counts. This range was selected to determine the I_{rms} distribution in the near-wall region. The results show substantial differences from the baseline in Section 3.1, which has a stable laminar state that lasts until x = 170 mm. The boundary layer upstream of the actuators is in the laminar state and has a thickness of approximately 1 mm. As the SAPAs are activated, the disturbance of the downstream boundary layer increases significantly. A slight increase in the displacement thickness and the pulsation in the boundary layer is observed, unlike in the raw schlieren images. It is likely caused by the consecutive pass-throughs of the TGBs. In **Figure 10**, the black bar at x = 80 mm downstream of the actuators indicates where the near-wall region has a much higher I_{rms} (> 1.6) than in the incoming laminar boundary layer and the downstream region. These peak I_{rms} contours occur from x = 85 mm to x = 130 mm, corresponding to the stronger fluctuation in the density gradient, indicating the formation of a shear layer with higher velocity fluctuations after the actuators have been

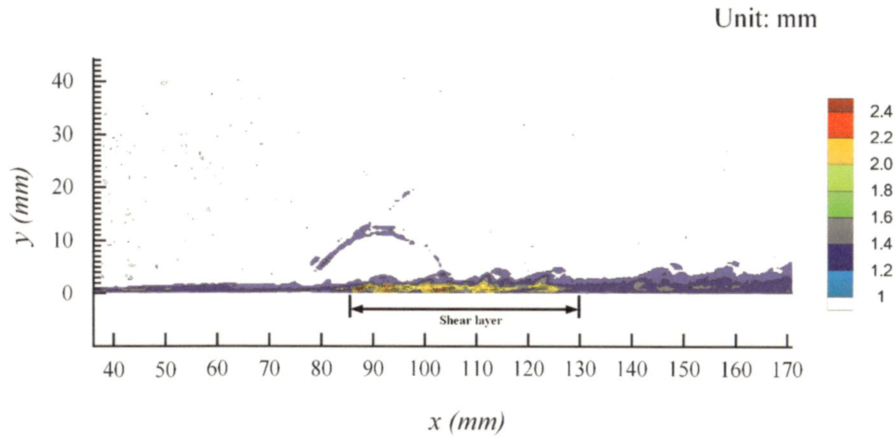

Figure 16.
The RMS of the Schlieren intensity field (I_{rms}) *for 10 kHz reputation actuation.*

activated. A closer inspection of the I_{rms} value in the entire range of the shear layer shows a small subregion close to the surface where the I_{rms} values are noticeably lower than in the above region, namely the separated shear layer. The thickness of the boundary layer is greater downstream of the shear layer as the turbulence increases at approximately x = 140 in agreement with the raw schlieren images. The observed increases in the boundary layer thickness and the I_{rms} values indicate that the transition zone has moved upstream after the activation of the SAPAs. The semiquantitative schlieren results confirm the excellent performance of the SAPAs for accelerating the transition by approximately 30 mm in this case. It should be noted that when SAPAs are used to trip the flow, an adequate distance must be chosen to allow the formation of vortices to impact the boundary layer. This distance is crucial. If the SAPAs are placed too far ahead, the laminar flow region is reduced, which increases the drag. Placing the devices far too close to the desired transition location renders them ineffective.

The above content demonstrated that surface arc discharge led to the formation of TGBs and streamwise trailing vortices in the boundary layer, strongly affecting its stability and inducing laminar-turbulent transition. The I_{rms} values for different actuation frequencies (20, 40 kHz) are compared in **Figure 17**. Due to the interaction with the TGBs, the near-wall region downstream of the actuators has substantially higher I_{rms} values than the region of the incoming laminar boundary layer, which agrees with the 10 kHz reputation actuation results in **Figure 16**, namely the shear layer. Its streamwise movement increases as the actuation frequency increases. A comparison of the locations with high I_{rms} values reveals that high-frequency actuation results in a short shear layer in the streamwise direction. Thus, a faster transition process occurs at higher actuation frequencies. Moreover, the peak I_{rms} for 20 and 40 kHz actuation exceeds 10, whereas that for 10 kHz actuation (**Figure 16**) is about 2.4. The increase in the peak I_{rms} suggests that strong density fluctuations occur downstream of the actuators. In compressible flow, the velocity and density fluctuations are correlated; thus, increased velocity fluctuations are expected in this region. As plasma actuation occurs, numerous periodic streamwise vortices are produced, especially at high-frequency actuation, resulting in an enhancement in the mixture and the momentum transfer in the boundary layer. This explains why the transition to turbulent is relatively fast at high frequencies.

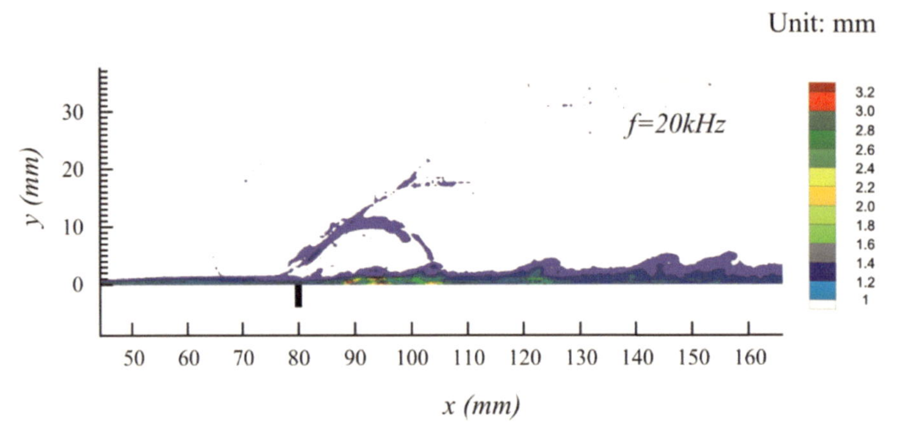

(a)

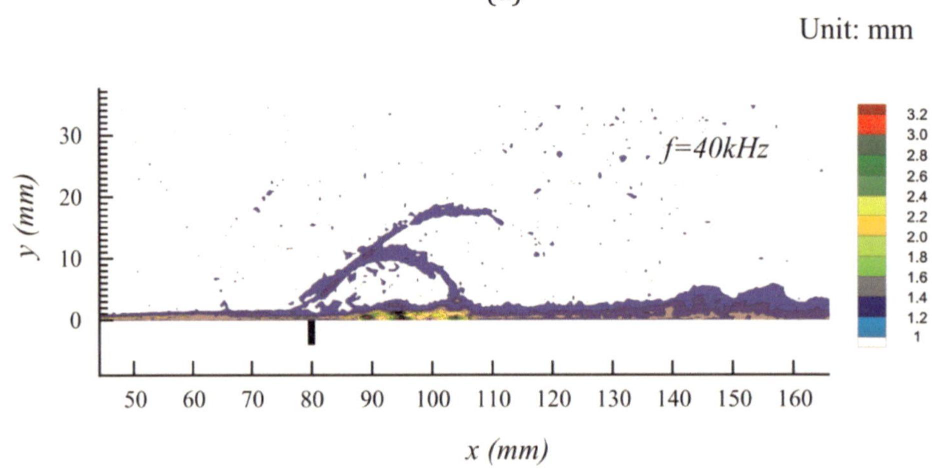

(b)

Figure 17.
Contours of I_{rms} *for boundary layer development with different actuation frequencies (a)* $f = 20$ *kHz (b)* $f = 40$ *kHz.*

5. Summary

An experimental study was performed to explore the effects of a SAPA array on boundary-layer transition on a flat plate. High-speed schlieren images and high-frequency pressure measurements were obtained. The RMS results of the schlieren intensity field and dynamic pressure yielded valuable insights into the ability of the SAPAs to induce laminar-turbulent transition.

A natural laminar to the turbulent transition of a supersonic boundary layer on a flat plate for Ma = 3 flow was investigated using the schlieren system with a minimal exposure time and a high image acquisition rate. The results indicated that the SAPAs achieved laminar to turbulent transition using a high repetition rate. The pressure spectra of the locations downstream of the actuators had the same magnitude of the characteristic frequency of the turbulent boundary layer as in the baseline condition, indicating that a transition from laminar to turbulent flow had occurred. In addition, the frequency range widened because many streamwise vortices were present in the boundary layer while the actuators were operating. Two higher actuation frequency modes were compared to determine the frequency effect on the laminar-turbulent transition. The I_{rms} results indicated that the transition location moved upstream as the actuation frequency increased; however, further solid evidence through direct measurements is needed.

Although we captured the laminar-turbulent transition of a supersonic boundary layer using high-frequency SAPAs, the pressure measurements and schlieren observations did not provide sufficient insights into the velocity profile and the scaling properties of the boundary layer. In the future, we plan a parametric investigation of the effects of the velocity characteristics. Particle image velocimetry could be used to investigate these flow fields in detail. In addition, the delay of the laminar boundary layer control by the plasma actuators should also need to be concerned.

Acknowledgements

The authors gratefully acknowledge the National Natural Science Foundation of China (NSFC) and China Postdoctoral Science Foundation (CPSF) for the financial support under Grant No. 11902360 and Grant No. 2021 M702676.

Author details

Gan Tian
Northwestern Polytechnical University, Xi'an, China

*Address all correspondence to: gantian_nwpu@163.com

References

[1] Alexander F. Transition and stability of high-speed boundary layers. Annual Review of Fluid Mechanics. 2011;**43**: 79-95. DOI: 10.1146/annurev-fluid-122109-160750

[2] Dolling DS. Fifty years of shock-wave/boundary-layer interaction research: What next? AIAA Journal. 2001;**39**(8):1517-1531. DOI: 10.2514/2.1476

[3] Saric WS, Reed HL, Kerschen EJ. Boundary-layer receptivity to freestream disturbances. Annual Review of Fluid Mechanics. 2002;**34**:291-319. DOI: 10.1146/annurev.fluid.34.082701.161921

[4] Wu Y, Li Y. Research progress and outlook of flow control and combustion control using plasma actuation. Scientia Sinica Technologica. 2020;**50**(10): 1252-1273. DOI: 10.1360/SST-2020-0111

[5] Samimy M, Kim JH, Kastner J, Adamovich I, Utkin Y. Active control of high-speed and high-Reynolds-number jets using plasma actuators. Journal of Fluid Mechanics. 2007;**578**:305-330. DOI: 10.1017/S0022112007004867

[6] Sato M, Nonomura T, Okada K, Asada K, Aono H, Yakeno A, et al. Mechanisms for laminar separated-flow control using dielectric-barrier-discharge plasma actuator at low Reynolds number. Physics of Fluids. 2015;**27**(11):487-545. DOI: 10.1063/1.4935357

[7] Zong H, Kotsonis M. Formation, evolution and scaling of plasma synthetic jets. Journal of Fluid Mechanics. 2018; **837**:147-181. DOI: 10.1017/jfm.2017.855

[8] Samimy M, Adamovich I, Webb B, Kastner J, Hileman J, Keshav S, et al. Development and characterization of plasma actuators for high-speed jet control. Experiments in Fluids. 2004; **37**(4):577-588. DOI: 10.1007/s00348-004-0854-7

[9] Shang JS, Surzhikov ST, Kimmel R, Gaitonde D, Menart J, Hayes J. Mechanisms of plasma actuators for hypersonic flow control. Progress in Aerospace Sciences. 2005;**41**(8): 642-668. DOI: 10.1016/j.paerosci.2005.11.001

[10] Leonov SB, Yarantsev DA. Near-surface electrical discharge in supersonic airflow: Properties and flow control. Journal of Propulsion and Power. 2008; **24**(6):1168-1181. DOI: 10.2514/1.24585

[11] Chernyshev SL, Kiselev AP, Kuryachii AP. Laminar flow control research at TsAGI: Past and present. Progress in Aerospace Sciences. 2011;**47**: 169-185. DOI: 10.1016/j.paerosci.2010.11.001

[12] Philipp CD, Markus JK. Crossflow transition control by upstream flow deformation using plasma actuators. Journal of Applied Physics. 2017;**121**(6): 063303. DOI: 10.1063/1.4975791

[13] Duchmann A, Simon B, Tropea C, Grundmann S. Dielectric barrier discharge plasma actuators for in-flight transition delay. AIAA Journal. 2014; **52**(2):358-367. DOI: 10.2514/1.J052485

[14] Grundmann S, Tropea C. Experimental transition delay using glow-discharge plasma actuators. Experiments in Fluids. 2017;**42**(4): 653-657. DOI: 10.1007/s00348-007-0256-8

[15] Yan H, Liu F, Xu J, Xue Y. Study of oblique shock wave control by surface

arc discharge plasma. AIAA Journal. 2017;**56**(2):532-541. DOI: 10.2514/1. J056107

[16] Webb N, Clifford C, Samimy M. Control of oblique shock wave/boundary layer interaction using plasma actuators. Experiments in Fluids. 2013;**54**(6):1545

[17] Zhang Z, Wu Y, Sun Z, Song H, Jia M, Zong H, et al. Experimental research on multichannel discharge circuit and multi-electrode plasma synthetic jet actuator. Journal of Physics D: Applied Physics. 2017;**50** (16):165205. DOI: 10.1088/1361-6463/aa6372

[18] Matsubara M, Alfredsson PH. Disturbance growth in boundary layers subjected to free-stream turbulence. Journal of Fluid Mechanics. 2001;**430**: 149-168. DOI: 10.1017/S0022112000002810

[19] Westin KJA, Bakchinov AA, Kozlov VV, Alfredsson PH. Experiments on localized disturbances in a flat plate boundary layer. Pt I. The receptivity and evolution of a localized free stream disturbances. European Journal of Mechanics - B/Fluids. 1998;**17**: 823-846. DOI: 10.1016/S0997-7546(99)80016-8

[20] Hong Y, Gaitonde D. Effect of thermally induced perturbation in supersonic boundary layers. Physics of Fluids. 2010;**22**(6):064101. DOI: 10.1063/1.3432513

[21] Tian G, Yun W, Sun ZZ, Jin D, Huimin S, Min J. Shock wave boundary layer interaction controlled by surface arc plasma actuators. Physics of Fluids. 2018;**30**(5):055107. DOI: 10.2514/1. J058257

[22] Gan T, Jin D, Guo S, Wu Y, Li Y. Influence of ambient pressure on the performance of an arc discharge plasma actuator. Contributions to Plasma Physics. 2018;**58**(4):260-268. DOI: 10.1002/ctpp.201700133

Section 3

Verification and Applications of Flow Mechanics

Chapter 5

Falkner-Skan Experimental Verification Failure for Airflow on a Wedge

David Weyburne

Abstract

Velocity profiles along a wedge obtained by computational fluid dynamic simulation of laminar airflow are compared to the Falkner-Skan solution to boundary layer flow. The experimental flow direction velocity profiles appear to show a good match to the theoretical predictions but, on closer examination, the comparison fails. The results also indicate that the Falkner-Skan scaled experimental normal velocity and the scaled normal pressure gradient profiles do not show profile similarity for the laminar airflow on a wedge. The one exception that does show good profile similarity for both the velocities and the pressure gradient is when the flow direction pressure gradient is zero, i.e. the Blasius flat plate flow model. Examining various profile plots indicates that the failure is due to a non-constant pressure gradient along the wedge wall that fundamentally changes the shape of the profiles at the different stations along the wall. This makes airflow velocity profile similarity impossible.

Keywords: Falkner-Skan, boundary layer flow, verification failure, airflow on a wedge, computational fluid dynamic simulation

1. Introduction

Since the introduction of the Falkner and Skan [1] paper in the 1930s, there have been thousands of journal articles written on different aspects of the Falkner-Skan solution to boundary layer flow (a Google Scholar search in May 2023 yielded 9800 article results). Along with the Blasius solution, it is covered in virtually every fluid flow textbook in existence. However, a search of the literature reveals the Falkner-Skan theoretical solution has never been verified experimentally. The fact that the Falkner-Skan solution has not been experimentally verified is surprising given that Falkner-Skan-designed wind tunnel experiments and measurements of the velocity profiles are widespread and even happen as part of undergraduate fluids courses.

It is perhaps the case that the flow community has felt that experimental verification was unnecessary given the routine nature of the measurements and the apparent agreement with Falkner-Skan theory. However, a closer look reveals that one reason for the lack of verification is due to the limitations of wind tunnel measurement capabilities. Velocity measurement noise limits high accuracy and pressure measurement probe size prevents measurements inside of the boundary layer from even being made. Furthermore, the normal velocity in a boundary layer is typically orders of magnitude smaller than the main flow which makes measurements of the normal velocity profile all but impossible. This means even with careful measurements of the velocity profile in the flow direction, the comparison to the Falkner-Skan theory would not be complete or fully definitive.

As a consequence of the wind tunnel difficulties, for the work herein, we turn to computational fluid dynamic (CFD) simulations of laminar flow on a wedge. The wedge simulation is chosen since the Falkner-Skan formulation is often identified as flow along a wedge in the literature, and in any case, represents a simple geometry that produces a pressure gradient in the flow direction along the wedge plate. Due to the simplicity of the flow model, we assume that the CFD simulations for laminar flow on a wedge are representative of the real-world situation without validation.

2. A Falkner-Skan review

Before proceeding to the simulation section, we first review the Falkner-Skan model for boundary layer flow. The Falkner and Skan [1] solution describes the steady boundary layer that forms for 2D laminar flow along a flat wall in which a pressure gradient is imposed in the flow direction. The Falkner-Skan theoretical approach to solving the boundary layer equations employs a series of key assumptions. The first key assumption, one that underlies the whole Falkner-Skan approach, is that there is a set of normal-to-the-wall velocity profiles at different stations along the wall that appear similar when scaled with a properly chosen length and velocity scaling parameters. That is, there must be a region along the wall in which plots of the scaled profiles collapse to a single curve. The velocity profile is defined as a series of velocity values taken from the wall out to the free stream in a direction normal to the wall. This non-dimensionalization process reduces the set of partial differential flow governing equations to an ordinary differential equation greatly simplifying the solution process.

Falkner-Skan employed a stream function approach to describe the velocity in the flow direction, $u(x,y)$, and the normal velocity, $v(x,y)$. The Falkner-Skan second key assumption is that a stream function $\Psi(x,y)$ exists (see [2], p. 543) such that

$$\psi(x,y) = \delta_s(x)\, u_s(x) f(\eta), \tag{1}$$

where $f(\eta)$ is a dimensionless function that only depends on the scaled y-position, $\eta = y/\delta_s(x)$, where $\delta_s(x)$ is the length scaling parameter, and where $u_s(x)$ is the velocity scaling parameter. The flow direction is along the wall in the x-direction with the y-direction as the normal to the wall direction. The stream function must satisfy the conditions

$$u(x,y) = \frac{\partial\psi(x,y)}{\partial y}, \quad \text{and} \quad v(x,y) = -\frac{\partial\psi(x,y)}{\partial x}. \tag{2}$$

The third key assumption Falkner-Skan made was that the scaling parameters $\delta_s(x)$ and $u_s(x)$ are simple power-law functions of the distance along the wall given by.

$$u_s(x) = a\left(\frac{x}{L}\right)^m \quad \text{and} \quad \delta_s(x) = b\left(\frac{x}{L}\right)^n, \tag{3}$$

where a, b, L, m, and n are constants. The constant L is the model reference length, e.g. the wedge length for the case where a wedge is used to induce the non-zero pressure gradient.

The fourth key assumption Falkner-Skan made was that Prandtl momentum equations apply rather than the full Navier-Stokes versions. Prandtl [3] used an order of magnitude argument to reduce the full Navier-Stokes momentum equations to a much simpler set of momentum equations known as the Prandtl momentum equations. Using these dimensional arguments, Falkner-Skan was then able to obtain velocity solutions using just the stream function based Prandtl x momentum conservation equation since the assumed y momentum contribution is negligible.

The Falkner-Skan fifth key assumption is related to the pressure gradient term in the Prandtl x-momentum conservation equation. Falkner-Skan assumed that the pressure gradient term can be replaced by the differential form of the Bernoulli equation in the high Reynolds number limit. This takes the form

$$-\frac{1}{\rho}\frac{\partial P(x,y)}{\partial x} = u_e(x)\frac{du_e(x)}{dx}. \tag{4}$$

Implicit in this substitution is that Falkner-Skan assumed the velocity scaling parameter $u_s(x)$ is given by the velocity at the boundary layer edge $u_e(x)$.

With the above assumptions, it is possible to reduce the boundary layer x-momentum equation to a simple ordinary differential equation that can be solved for $f(\eta)$ and its derivatives by standard differential equation techniques. The resulting expressions for the $u(x,y)$ profile, the scaled $v(x,y)$ profiles, and the scaled y-pressure gradient profile are given by x-momentum by

$$\begin{aligned} u(x,y) &= u_s f'(\eta), \\ v(x,y) &= \frac{u_s\delta_s}{x}\left[-\frac{1+m}{2}f(\eta) + \frac{1-m}{2}\eta f'(\eta)\right], \quad \text{and} \\ \frac{1}{\rho}\frac{dP}{dy} &= \frac{u_s^2\delta_s}{x^2}\begin{bmatrix} -\frac{1}{2}(3m-1)f''(\eta) + \frac{1}{2}(1-m)\eta f'''(\eta) - \frac{1}{4}(m+1)^2 f(\eta)f'(\eta) + \\ \frac{1}{4}(m-1)^2\eta f'^2(\eta) - \frac{1}{4}(m^2-1)\eta f(\eta)f''(\eta) \end{bmatrix}, \end{aligned} \tag{5}$$

where m is the power coefficient from Eq. (3) and the primes represent derivatives of $f(\eta)$ with respect to η. The y-pressure gradient expression was developed by substituting the Eq. (5) velocity expressions into the y-momentum conservation equation [4]. Eq. (5) represents the velocity and pressure gradient expressions that will be compared to the CFD simulation results.

3. Experimental design

The Falkner-Skan experimental verification process is based on OpenFOAM CFD simulations of 2D exterior-like laminar airflow along a wedge. The depiction of the simulation geometry is shown in **Figure 1**. The bottom solid line is the wedge wall surface of length L and has a wall boundary condition. The top and bottom black dashed lines are assigned a zero gradient boundary condition. The bottom dashed black line is the inlet region of length L_{in}. The flow inlet and outlet regions, height H,

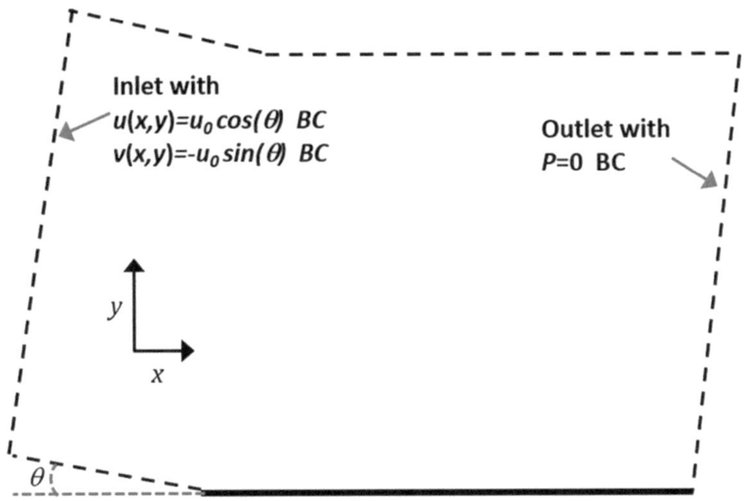

Figure 1.
Depiction of the 2D laminar flow simulation geometry on a wedge.

are slanted θ degrees to the wall normal. The velocity in the x-direction is desigmated $u(x,y)$ and the velocity in the y-direction is designated $v(x,y)$. The flat wedge geometry utilizes a 3300 × 1000 mesh that emphasizes the near wall and leading-edge regions. A total of five simulations are performed at five wedge angles, $\theta = -5°, 0°, 5°, 10°$, and 20°. The inlet RT airflow is set to $u_0 = 0.625\ m/s$ resulting in a $Re_x = 5.0 \times 10^5$ at the end of the 12 m long wedge wall. The low velocity enabled the easier-to-converge incompressible option to be used. The simulation extents are based on a series of preliminary test cases. The height extent value is set so that exterior-like flow is generated as judged by the asymptotic approach of the normal velocity profile to zero in the free stream region above the wall. The inlet location was increased until there were minimal changes in the boundary layer along the wall. This resulted in the inlet (bottom black dashed line) being set to $L_{in} = 80\ m$ and the height being set to $H = 500\ m$. The simulations are iterated until the velocity and pressure residuals all drop below a value of 1×10^{-6}. To be consistent with the Falkner-Skan approach, the velocity and pressure profiles are measured normal to the wall.

4. Parameter extraction

An example of the simulation result is shown in **Figure 2** for the of $\theta = 10°$ wedge angle case. Each velocity profile consists of a series of velocity values taken from a point on the wedge surface to a point deep into the free stream in a direction normal to the wedge plate wall (y-direction). u_0 is the inlet velocity and L is the wedge length. The experimental $u(x,y)$ velocity profiles shown in **Figure 2a** show a much more nuanced boundary layer picture than is typically discussed in the literature. The question becomes what part of this complicated boundary layer picture corresponds to the Falkner-Skan boundary layer model. Falkner and Skan made it clear in their paper that the formulation is meant to describe the near-wall portion of the profiles where viscosity is important. Examination of **Figure 2a** reveals that after the initial jump from $u(x,0) = 0$, the profiles all asymptote to a velocity plateau near $y/L = \sim 0.01$. This velocity corresponds to the velocity at the viscous boundary layer edge, $u_e(x)$. For reproducibility, the actual $u_e(x)$ values are extracted using a method based on

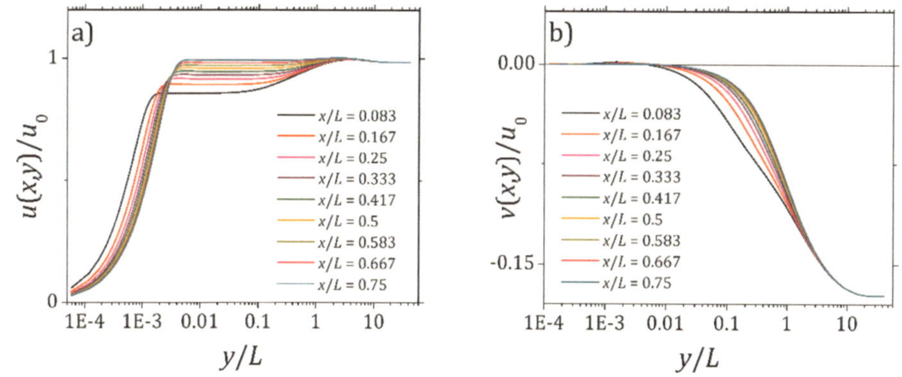

Figure 2.
(a) simulated 2D laminar airflow velocity profiles for u(x,y) *at nine locations along the wedge plate for the* θ =10° *wedge. (b) nine normal velocity profiles,* v(x,y), *at the same locations.*

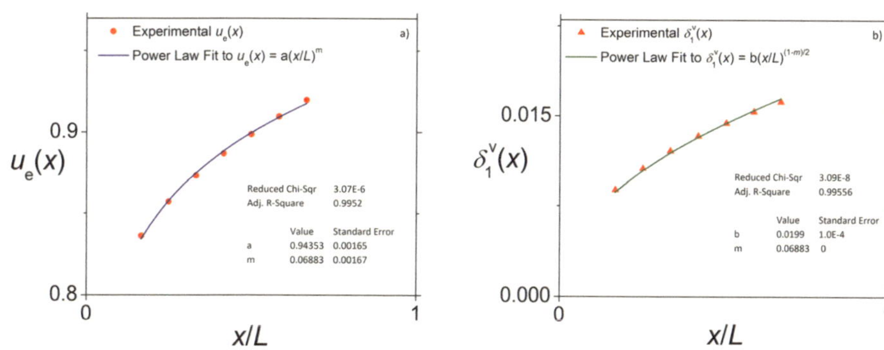

Figure 3.
The red circles are the extracted boundary layer edge velocity values, $u_e(x)$, *at seven locations along the wedge plate and the blue line is the power law fitted result.*

calculating the viscous thickness. In this interpretation, the scaling parameter $u_e(x)$ is given the value of the velocity, $u(x,y)$, evaluated at $y = \delta_V(x)$, where $\delta_V(x)$ is the four-sigma viscous thickness. The four-sigma $\delta_V(x)$ location is calculated by the integral moment method [5]. It corresponds to the place where the majority of the viscous momentum contributions have vanished. The four-sigma $\delta_V(x)$ calculated location values ranged from $y/L = 0.06$ to $y/L = 0.1$ corresponding to the plateau regions in **Figure 2a**.

The extracted $u_e(x)$ values are plotted in **Figure 3** as red circles for the $\theta = 10°$ case. The seven-profile subset $x/L = 0.25$ to $x/L = 0.75$ is chosen based on this set yielding the power law fit with the smallest adjusted R-square value. The blue line shows the power law fit, Eq. (3), with an extracted power-law fitted exponent, $m = 0.07253$, for this case. The extraction process for the $\theta = 0°$ Blasius case determined that, as expected, the fit to $m \cong 0$ is very good for this case.

With the $u_e(x)$ power law fitting parameter in hand, the next step is to extract the relevant fitting parameters for the length scaling parameter $\delta_s(x)$ in Eq. (3). This parameter was never identified in the original Falkner-Skan article. However, to reduce the x-momentum equation to a simple differential equation, Falkner-Skan showed that power-law exponents in Eq. (3) are not independent but must satisfy the equation given by that

$$m + 2n + 1 = 1, \quad (6)$$

using the fitted m value from **Figure 3**, then for the θ = 10° case, n = 0.463735. The b parameter value is not fixed by the theory and can therefore be adjusted for best fit between the Falkner-Skan solution and the scaled experimental profiles.

5. Profile comparison

With the scaling values determined, the next step of the verification process is to verify profile similarity. This is accomplished by plotting the experimental scaled $u(x,y)$ profiles, the scaled $v(x,y)$ profiles, and the scaled y-pressure gradient profiles and comparing these to the Falkner-Skan theoretical profiles given by Eq. (5). Falkner-Skan verification confirmation requires that the set of experimental profiles plotted as $y/\delta_s(x)$ versus $u(x,y)/u_s(x)$, for example, should all collapse to a single curve corresponding to the profile given in Eq. (5). In **Figures 4–6**, the results for the θ = 10° case and the θ = 0° cases are shown. In these figures, the Eq. (3) fitted values to the scaling parameter $u_e(x)$ are used. The results for the θ = −5°, θ = 5°, and the θ = 20° simulations are not shown but show similar behavior to those in **Figures 4a–6a**. In each case, the Falkner-Skan $f(\eta)$ and its derivatives are calculated with a standard Fortran program. The b parameter is adjusted to make the Falkner-Skan $u(x,y)$ velocity and the experimental scaled profiles overlap in **Figure 4a**.

The Falkner-Skan scaled $u(x,y)$ profiles in **Figure 4a** show reasonably good overlap with the theoretical curve whereas the **Figure 4b** Blasius θ = 0° case shows excellent overlap. However, a closer look at the **Figure 4a** result tells a different story. In **Figure 7** we zoom in on a small region of the velocity profile curves from **Figure 4**. The θ = 10° curves no longer overlap at the expanded scale indicating profile similarity is absent.

The normal velocity and the y-pressure gradient experimental profiles, **Figures 5a** and **6a**, also do not exhibit profile similarity and do not collapse to the Falkner-Skan

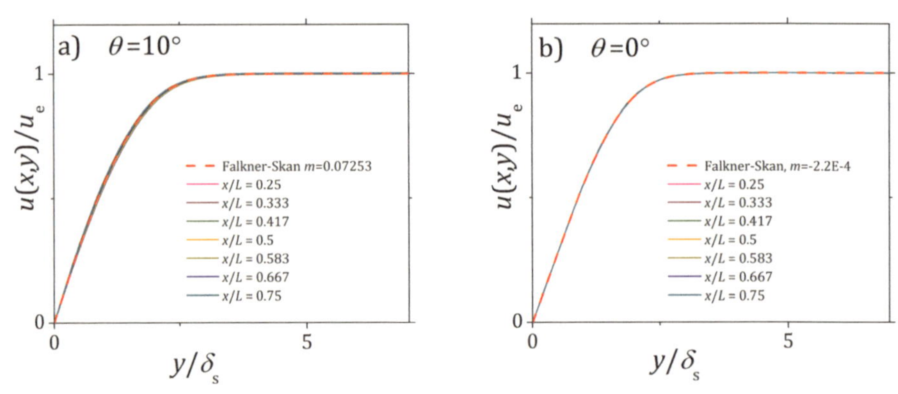

Figure 4.
(a) scaled u(x,y) *velocity profiles at seven locations along the wedge plate for the* θ=10° *case, (b) scaled velocity profiles at seven locations along the wedge plate for the* θ=0° *Blasius case. The dashed red line is the Falkner-Skan profile.*

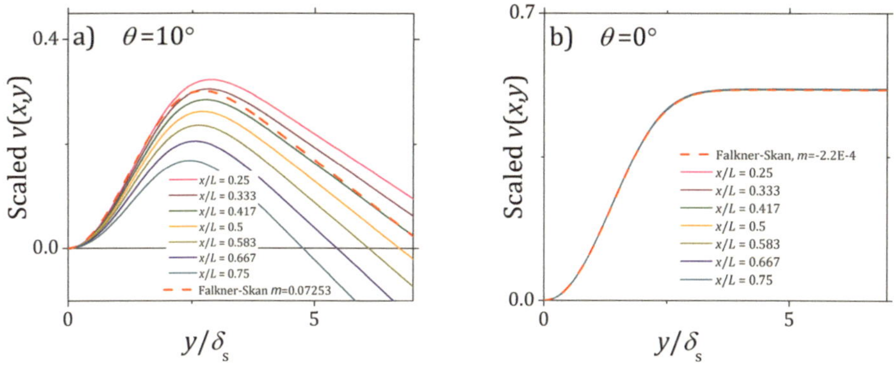

Figure 5.
(a) scaled v(x,y) *velocity profiles at seven locations along the wedge for the* θ=10° *case, (b) scaled velocity profiles along the wedge plate for the* θ=0° *case. The dashed red line is the Falkner-Skan profile and the velocity scale is* $\delta_s u_s/x$.

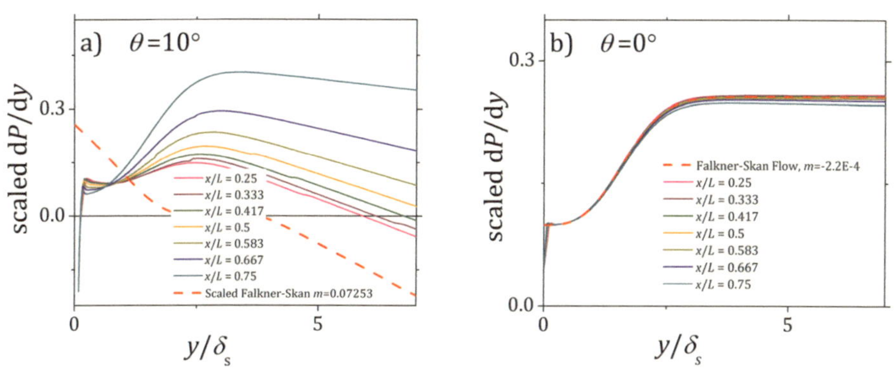

Figure 6.
(a) scaled dP/dy *profiles at seven locations along the wedge plate for the* θ=10° *case, (b), the scaled* dP/dy *profiles along the wedge plate for the* θ=0° *case. The dashed red line is the Falkner-Skan profile and the* y*-scale is* $\delta_s u_s^2/x^2$.

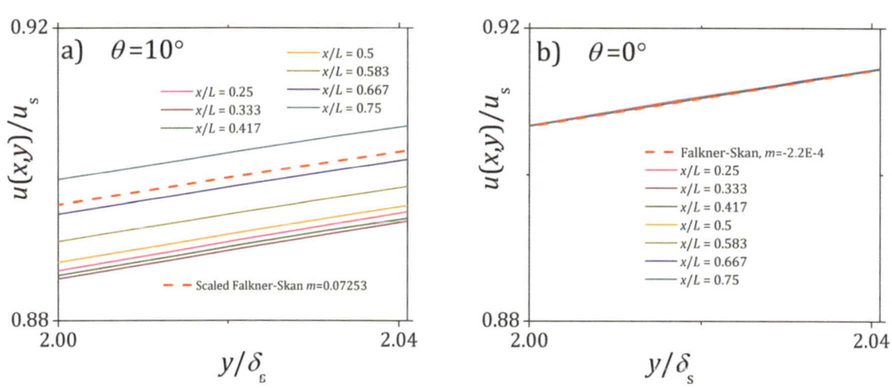

Figure 7.
(a) a zoomed-in plot of the scaled u(x,y) *velocity profiles from* ***Figure 4a*** *for the* θ = 10° *case, (b) a zoomed-in plot of the scaled* u(x,y) *velocity profiles from* ***Figure 4b*** *for the* θ = 0° *case.*

profiles. All of the non-zero angle cases show the same type of non-similar behavior demonstrated in **Figures 5a, 6a** and **7a**. The conclusion is that: 1) although the Falkner-Skan solution for laminar airflow does a reasonable job of approximating the $u(x,y)$ profiles, it fails upon closer examination and 2) the normal velocity $v(x,y)$, and the y-pressure gradient profiles do not show profile similarity using the Falkner-Skan scaling parameters.

6. The explanation

The key to understanding the nature of the failure is the one confounding case that does show good profile similarity for both the velocity and pressure gradients. When the flow direction pressure gradient is zero, i.e. the $m = 0$ Blasius flow model, profile similarity is obtained as shown in **Figures 4b**, **5b**, and **6b**. The fact that the similarity failure occurs for the non-zero-pressure gradient cases but similarity does occur for the zero-pressure gradient case implicates the flow direction pressure gradient as the problem area.

One possibility is that the power law-based Bernoulli equation approximation consisting of Eqs. (3) and (4) is a problem area. An examination of the scaled profiles reveals that the only way to force similar behavior is by allowing the Falkner-Skan $\delta_s(x)$ and $u_s(x)$ scaling parameters a and b constants to change with x. This compromises the whole attempt to reduce the partial differential flow governing equations to the Falkner-Skan ODE. Even if one ignores the Falkner-Skan $\delta_s(x)$ and $u_s(x)$ scaling parameters and tries different scalings to try to force similar behavior (for example V_{max} and y at V_{max} for the velocity and length scaling in **Figure 5a**), one discovers that the problem is that is no simple x and y shifts that can result in the velocity profiles or the y-pressure gradient profiles at different stations to overlap completely. The profiles have a different shape at each station along the wall!

Since the Blasius case does show similarity and the non-zero angle cases do not means that the x-pressure gradient is affecting the shape of the profile. What sets the x-pressure gradients apart for these cases? The x-pressure gradients for all five test cases are plotted along the wedge wall just above the viscous boundary layer in **Figure 8**. One of the clear observable differences is that only the $\theta = 0°$ case is showing a constant pressure gradient along the wall. This becomes important when one considers that the relationship between the second derivative of the velocity at the wall and the x-pressure gradient is given by the Prandtl x-momentum equation evaluated at the wall ($y = 0$) as

$$v\frac{\partial^2 u(x,0)}{\partial y^2} = \frac{1}{\rho}\frac{\partial P(x)}{\partial x} \quad . \tag{7}$$

This means that the wall value of the second derivative of the velocity can vary from a negative value to a positive value depending on whether we have an adverse pressure gradient or a favorable pressure gradient. The x-pressure gradient in the y-direction of the boundary layer region is known to be mostly constant. Hence, the non-constant pressure gradients observed at the boundary layer edge is changing the shape of the second derivative of the velocity as we move along the wall. This is illustrated in **Figure 9** in which the second derivatives of the $u(x,y)$ velocity profiles from **Figure 4** are plotted for the $\theta = 10°$ and $\theta = 0°$ cases. No simple x and y scaling will make the

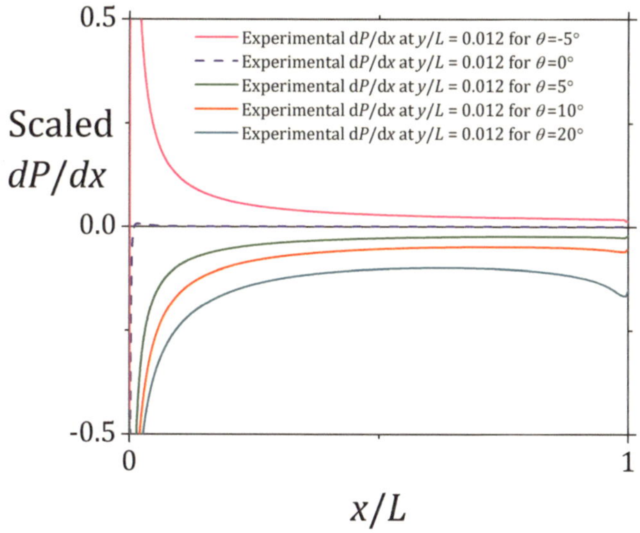

Figure 8.
The experimental x-*pressure gradient curves for the five simulation cases measured at* y/L = 0.012 *along the wedge surface which is just above the viscous boundary layer thickness.*

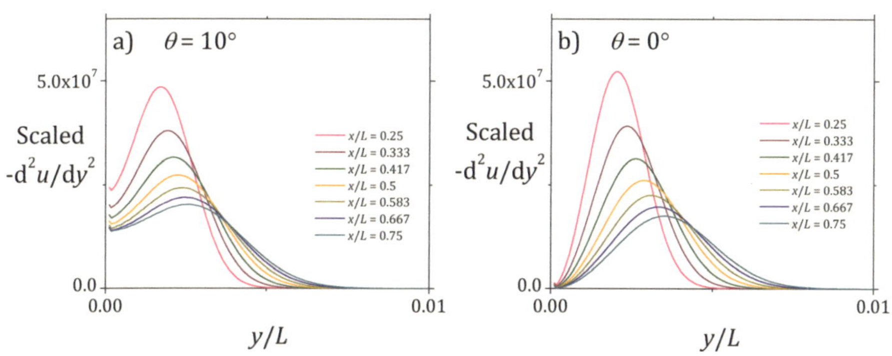

Figure 9.
(a) experimental scaled second derivative profiles for the θ=10° *velocity profiles from* **Figure 2**, *(b) the scaled second derivative profiles for the* θ=0° *Blasius case. The* y-*scale in both plots is* u_o/L^2

θ = 10° profiles look similar whereas the θ = 0° profiles look Gaussian-like and visually appear to have a similar shape. If the second derivative profile shapes are different then velocity profiles obtained by twice integrating will be different. This is what leads to the non-overlapping behavior evident in **Figure** 7. The non-constant x-pressure gradient inducing profile changes explains all the behavior observed in **Figures** 4–7. The first key assumption of the Falkner-Skan model is that there must be a region along the wall where the shape of the profiles from different stations are the same. This is not the case for CFD laminar airflow cases on a wedge observed in this effort.

7. Falkner-Skan is not wedge flow

The Falkner-Skan formulation is often identified as fluid flow along a wedge in the literature. The reason is that the potential flow solution to flow on a wedge appears similar to the outer flow predicted by the Falkner-Skan formulation (see [6], p. 156). It is certainly true that the wedge induces a pressure gradient in the flow direction along the wedge surface. However, the assumption in the literature has been that the viscous boundary layer shape remains the same for any induced pressure gradient. What is new here is the realization that laminar flow on a finite-sized wedge does not induce a constant pressure gradient along the wall and that the profile shape along the wall changes with a changing pressure gradient.

The trend lines in **Figure 8** indicate that the pressure gradient changes become larger along the wedge as the wedge angle becomes larger but the changes are observable at even small angles. The leading edge induced pressure changes are expected to occur even if one moves the exit past the wedge endpoint. We note that the geometry used in this report is a geometry that is at least possible to mimic in a wind tunnel. Therefore, the failure cannot be attributed to the choice of the wedge angle or that the simulations are not representative of a realizable flow situation.

It appears that boundary layer similarity for airflow is limited to cases where the pressure gradient is constant. At present, the only case that can be experimentally generated that satisfies that condition is the unbounded exterior-like ZPG flow on a flat plate case. Other constant pressure gradient cases should also work but the problem is that it is not obvious how one would create a constant non-zero pressure gradient in a wind tunnel or in a simulation due to the leading-edge pressure bubble. We have verified that a thick flat plate also has a leading-edge pressure bubble but further along the wall, the pressure gradient becomes relatively constant. One geometry that is presently being explored is the case where the power law exponent in Eqs. (3) and (4) is $m = 0.5$. The Bernoulli equation approximation for the pressure gradient in Eq. (4), with $m = 0.5$ (half angle = 60°), would indicate the pressure gradient should be constant past the leading-edge pressure bubble.

This realization that the velocity profile shape along the wall changes with a changing pressure gradient laminar has implications for more than just the wedge flow case. There are other flow cases in the literature that assume profile similarity is possible where non-constant pressure gradients are present. Flow in a converging or diverging channel (sink flow) is an example. Thus, it is important that flow situations in the literature that assume the profile shape remains the same for cases in a non-zero pressure gradient need to be revisited.

8. Conclusions

Contrary to expectations, the simulated laminar airflow experiments along a wedge using computational fluid dynamic (CFD) do not compare well to the Falkner-Skan solution to boundary layer flow. While the experimental flow direction velocity profiles appear to show a good match to the theoretical predictions, on closer examination the comparison fails. The success of the one exception that does show good profile similarity for both velocities and the pressure gradient, i.e. the Blasius flat plate flow model, is attributed to having a constant pressure gradient in the flow direction. All other non-zero angle cases show non-constant pressure gradients. Theory, and the above experimental results, indicate that a non-constant pressure gradient along the wedge wall fundamentally changes the shape of the profiles at the different stations along the wall. This makes profile similarity impossible for airflow on a wedge.

DOI: http://dx.doi.org/10.5772/intechopen.1002293

Acknowledgements

The author acknowledges the support of the Air Force Research Laboratory and Gernot Pomrenke at AFOSR.

Author details

David Weyburne
Air Force Research Laboratory, Boston, USA

*Address all correspondence to: david.weyburne@gmail.com

References

[1] Falkner V, Skan S. Solutions of the boundary-layer equations. The London, Edinburgh, and Dublin Philosophical Magazine and Journal of Science. 1931; **12**:865-896. DOI: 10.1080/14786443109461870

[2] Panton R. Incompressible Flow. 4th ed. Hoboken, New Jersey: Wiley; 2013

[3] Prandtl L. *Über Flüssigkeitsbewegung* bei sehr kleiner Reibung. Verhandlungen es Dritten Internationalen Mathematiker-Kongresses in Heidelberg 1904. 1905;**2**:484-491

[4] Weyburne D. The Normal to the wall pressure gradient for Blasius and Falkner-Skan boundary layer flow. In: Weyburne D, editor. Aspects of Boundary Layer Theory. 1st ed. Maynard, Massachusetts: BasicScience; 2022

[5] Weyburne D. New thickness and shape parameters for the boundary layer velocity profile. Experimental Thermal and Fluid Science. 2014;**54**:22-28. DOI: 10.1016/j.expthermflusci.2014.01.008

[6] Schlichting H. Boundary-Layer Theory. 7th ed. New York, New York: McGraw-Hill; 1979

Chapter 6

The Influence of Flow Admixtures to the Electromagnetic Flow Meter Accuracy

R. Račkienė and J.A. Virbalis

Abstract

The measurement error appears in an electromagnetic flow meter when the flow has admixtures with magnetic and/or electric properties different from the fluid. The particle shape approximates the ellipsoid. The reasons of error appeared inside and outside of the particles in the active zone of the flow meter with any canal form and turbulent flow investigated. The investigation is performed by analogy among electrostatic, magnetic, and electric current fields. Expressions for error depending on volume concentration, permeability, electric conductivity of particles, and different ellipsoid axes lengths are presented. The expressions of the error are obtained assuming that particles are oriented in various directions concerning the flow direction. Magnetic particles with the considerable value of electric conductivity are dangerous in particular. The extra measurement error depends on particle shape in this case especially. The measurement error increases if the particle shape differs from the sphere. The complementary measurement error can exceed some hundred times the volume concentration of particles if the ratio between the longest and shortest axes of the ellipsoid exceeds 10. The error is proportional to the second power of admixtures volume concentration when particles are nonmagnetic. When admixtures are nonmagnetic and nonconductive, the measurement error does not depend on the particle's shape.

Keywords: electromagnetic flow meter, turbulent flow, admixtures, measurement error, ellipsoidal shape of particles, magnetic particles, conducting particles

1. Introduction

Electromagnetic flow meters (EMFM) for the measurement of ionic fluid flow in closed filled pipes are investigated in this chapter. Such meters are especially suitable in modern measuring systems, which are most often part of artificial intelligence. Their natural electrical output signal and absence of inertia allow you to create high-speed control and information systems.

The theory of the electromagnetic fluid flow meter was developed in the middle of the twentieth century. The Poisson equation describing the potential induced in the active zone of the electromagnetic flow meter, i.e., in that part of the meter where the magnetic field operates, was called by J. Shercliff [1] the main equation of the theory of electromagnetic flowmetry:

$$\nabla^2 V_i = \mathrm{div}[\boldsymbol{v} \times \boldsymbol{B}]. \tag{1}$$

This differential equation reveals the absence of measurement inertia. By solving it, we can find out how the induced potential V_i will be distributed in the active zone, if we know the distribution of the velocity of the liquid $\boldsymbol{v}$ and the magnetic flux density $\boldsymbol{B}$ in it. However, this equation is not tied to the design of the meter. In the case when the vector of the magetic flux density $\boldsymbol{B}$, the axis of the flow meter channel, and the straight line connecting the centers of the electrodes are perpendicular to each other, the induced voltage of the electrodes is convenient to express using the expression of electrode signal U_e proposed by M. Bevir [2]:

$$U_e = -\int_{\tau_a} (\boldsymbol{J}[\boldsymbol{v} \times \boldsymbol{B}])\mathrm{d}\tau = \int_{\tau_a} (\boldsymbol{v}[\boldsymbol{B} \times \boldsymbol{J}])\mathrm{d}\tau = \int_{\tau_a} \boldsymbol{v}\boldsymbol{W}\mathrm{d}\tau, \tag{2}$$

where τ_a is the volume of active zone of flow meter, $\boldsymbol{J}$ is the vector of virtual current density, coorespondingly, in any point of active zone, $\boldsymbol{W}$ is theweight vector. The virtual current density $\boldsymbol{J}$ is a formal parameter. It can be calculated as density of the current driving in the fluid at rest when to the electrodes of flow meter is connected source of current equal to 1 A [2]. Bevir proved this dependence as a theorem, so it accurately applies the basic equation of the electromagnetic flow meter to the selected design. The virtual current and weight vector are the fundamental units in theory of electromagnetic flow meters which allow to evaluate the influence of different parameters to the measurement accuracy. In [3], it is proposed to use the virtual current for the verification of the electromagnetic flow meters.

The liquid whose flow rate is measured is considered homogeneous. The theory of electromagnetic flow meters for homogeneous fluids has developed sufficiently. However, this is not always than we can suppose that liquid is homogeneous. Again the liquid may be contaminated with certain impurities or due to the leaking of the pipeline air bubbles can get into it. Therefore, it is important to find out how different admixtures change the measurement signal U_e.

With the development of the theory of electromagnetic flow meters, there were considered cases when the liquid is heterogeneous [4–8], but these were partial cases, usually intended at one type of heterogeneity. At Kaunas University of Technology, there were carried out systematic research to correlate the magnitude of measurement errors arising from various types of admixtures with the different physical properties and shape of those admixtures [9–12]. This chapter of the book summarizes the results obtained by Department of Electrical Power Systems of Kaunas University of Technology.

2. The basic equations, admixture shapes, and coordinate systems

Eq. (2) is especially suitable for the study of the influence of various admixtures on the electrode signal U_e, because the magnetic properties of admixtures will change the distribution of the magnetic flux density, and the electrical properties will change the distribution of virtual current, so we can perform a strict quantitative analysis.

The electrode signal of the electromagnetic flow meters is formed by any point of the active zone. We joined with the active zone the global rectangular coordinate system *xyz* (see **Figure 1**). The influence of any active zone point *x, y, z* to the measurement signal depends on the value of the weight vector in this point $\boldsymbol{W}(x,y,z) = \boldsymbol{B}(x,y,z) \times \boldsymbol{J}(x,y,z)$ [2].

The expressions for measurement signal errors when the fluid is contaminated by small magnetic particles depending on volume concentration, permeability, and electric conductivity of particles are obtained in [9] for spherical particles and for an ideal electromagnetic flow meter with a rectangular duct and infinitely conductive large electrodes.

The real shape of particles is different from the sphere frequently. The shape of particles will change the distribution of densities of virtual current and magnetic flux when the physical properties of particles are different than fluid. Therefore, it is important to investigate the dependence of measurement error on the shape of admixture particles.

We approximate the particle shape by an ellipsoid. This shape allows generalizing the particles of very different forms. The limiting cases of ellipsoids are spheres, disks, cylinders, lamella, and others. Some different particular cases are investigated in [10, 11]. We generalize all cases for the ellipsoidal shape of particles and any form of a channel with the point electrodes. We suppose that fluid flow is parallel to z axis, i.e., flow velocity has only component v_z, z axis coincides with channel axis, and x axis coincides with the line jointed centers of electrodes. The investigation is performed for flat flow velocity distribution in the channel cross section and the mean value of velocity equal to $\overline{v} = 1\,\text{m/s}$. We call this regime as normalized. Signal U can be expressed in this case [9]:

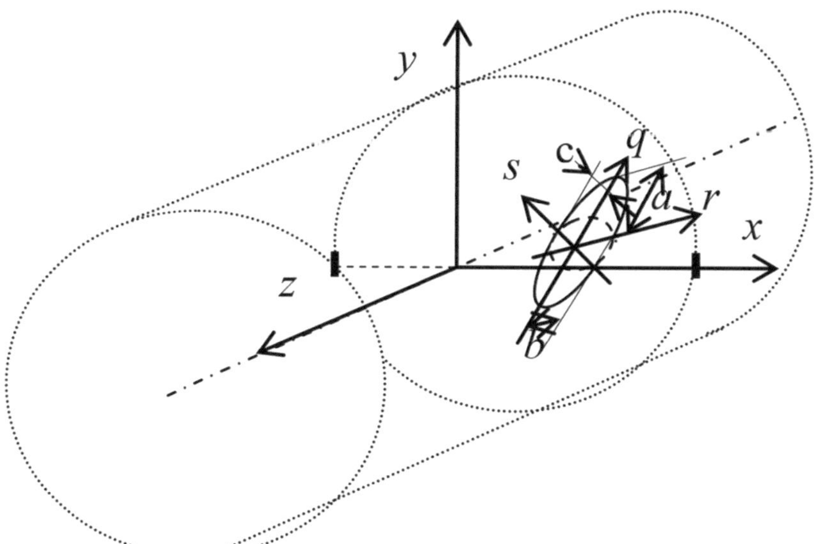

Figure 1.
Global and local coordinate systems.

$$\begin{cases} U = \int_{\tau_a} W_z \mathrm{d}\tau, \\ W_z = J_x B_y - J_y B_x. \end{cases} \tag{3}$$

where τ_a–volume of the active zone; $W_z = W_z(x, y, z)$–z component of the weight vector in the point x, y, z of the active zone, $J_x(x, y, z)$, $B_x = B_x(x, y, z)$, $J_y = J_y(x, y, z)$, $B_y = B_y(x, y, z)$–the x and y components of the virtual current $\boldsymbol{J}$ and magnetic flux $\boldsymbol{B}$ densities vectors in this point.

Using expression (3), we can evaluate how the magnetic and electric properties of the admixtures act to measurement error. To find the admixture influence to the electrode signal, it is needed to investigate the variation of the magnetic flux and the virtual current densities distribution in any active zone point. Evaluating that the weight vector can be different in the any active zone point we express the value of the measurement signal U_0 when the admixtures are absent in the flow as follows:

$$\begin{cases} U_0 = \left(\dfrac{1}{\tau_a} \displaystyle\int_{\tau_a} W_{z0} d\tau_a \right) \cdot \tau_a = \overline{W_{z0}} \tau_a, \\ \overline{W_{z0}} = \overline{J_{x0} B_{y0}} - \overline{J_{y0} B_{x0}}, \end{cases} \tag{4}$$

where $J_{x0} = J_{x0}(x, y, z)$, $J_{y0} = J_{y0}(x, y, z)$, $B_{x0} = B_{x0}(x, y, z)$, and $B_{y0} = B_{y0}(x, y, z)$–the values of the suitable components of the vectors $\boldsymbol{J}$ and $\boldsymbol{B}$ in the point x, y, z of the active zone without admixtures, $\overline{W_{z0}}$, $\overline{J_{x0} B_{y0}}$, and $\overline{J_{y0} B_{x0}}$ the mean values of the weight vector z component and the products of the suitable components of $\boldsymbol{J}$ and $\boldsymbol{B}$ in the clean fluid flow.

The shape of the admixture particle we approximate by an ellipsoid which orientation in respect to the global coordinate system can be any. We use the local rectangular coordinate system q, r, s whose axes coincide with the axes of the ellipsoid as it is shown in **Figure 1**. The equation of ellipsoid when the longest semi-axis a coincides with the q axis, the semi-axis of the middle length b coincides with the r axis and the shortest semi-axis c coincides with the s axis is:

$$q^2/a^2 + r^2/b^2 + s^2/c^2 = 1 \tag{5}$$

In this equation, a, b, and c are the lengths of the ellipsoid semi-axes. We can obtain very different forms of particles varying the ratios a/b and b/c.

In the general case, the measured flow can be turbulent or laminar. However, in the case of laminar flow, it is difficult to obtain generalizing expressions. Magnetic particles can be exposed to the magnetic field of the active zone, and they will begin to drift along the lines of the magnetic field, and that drift will depend on the flow rate. Heavier particles will be affected by the force of gravity, they will move to the bottom, and that movement will also depend on the speed of the liquid.

Therefore, we will limit further consideration to the turbulent flow. In the turbulent stream, fine particles will rotate and it can be assumed that they will not be affected by the gravity and magnetic forces.

It is very important to find out all the causes of errors, since they are also relevant when using an electromagnetic flow meter for measuring two-phase and three-phase flows.

In [9], four different components of signal error are indicated when some admixture particles get into active zone of the flow meter. The first component emerges due to the variation of virtual current and (or) magnetic flux density in the volume which the particles occupied. The other components arise for the variation of virtual current and magnetic flux density in the volume outside the particles. We can divide three different error components arising outside the admixture particles: because of the distortion of the virtual current and the magnetic flux density lines (supposing that its magnitudes were not varied), because of the variation of the magnetic flux density magnitude and because of the variation of the virtual current magnitude. Let's evaluate these.

3. The component of the signal error inside the admixture particle

The mean value of the weight function inside the ellipsoidal particle we can express this way:

$$\overline{W}_{zp} = \overline{J_{xp}B_{yp}} - \overline{J_{yp}B_{xp}} = K_F^I\overline{J_{x0}B_{y0}} - K_F^{II}\overline{J_{y0}B_{x0}}. \tag{6}$$

We suppose that particles are small and distributed evenly in the flow. The components $J_x = J_x\,(x, y, z)$, $J_y = J_y\,(x, y, z)$, $B_x = B_x\,(x, y, z)$, and $B_y = B_y(x, y, z)$ are not varied strictly by varying x, y, z. For small particles, we can suppose that independently on the meter design the components $J_x = J_x\,(x, y, z)$, $J_y = J_y\,(x, y, z)$, $B_x = B_x\,(x, y, z)$, and $B_y = B_y(x, y, z)$ have the same value in all volume which particle occupies before the particle gets into the flow, i.e., they are distributed uniformly. Therefore, in the ellipsoidal particle volume, they will be distributed uniformly too (see [13]).

We will study the distribution of the magnetic field of the electrostatic field in the dielectric ellipsoid, presented in [13], and the analogy of the electrostatic, magnetic, and electrical current fields.

Suppose that before the dielectric ellipsoid gets in field, there was a homogeneous electrostatic field $\boldsymbol{E}$, the components of which in the local coordinate system directed by the semi-axes of ellipsoid, a, b, and c were E_{a0}, E_{b0}, and E_{c0}. Inside the dielectric ellipsoid, the components by [13] will be

$$E_{ap} = \frac{C_a\varepsilon_f E_{a0}}{(C_a - 1)\varepsilon_f + \varepsilon_p}, E_{bp} = \frac{C_b\varepsilon_f E_{b0}}{(C_b - 1)\varepsilon_f + \varepsilon_p}, E_{cp} = \frac{C_c\varepsilon_f E_{c0}}{(C_c - 1)\varepsilon_f + \varepsilon_p}, \tag{7}$$

where ε_p is the relative permittivity of the ellipsoid and ε_f is the relative permittivity of the environment, C_a, C_b, and C_c are the ellipsoid shape factors expressed in [13] as follows:

$$\begin{cases} \boldsymbol{C_a} = \dfrac{2}{\boldsymbol{abc}\displaystyle\int_0^\infty \mathrm{d}w\Big/\sqrt{(w+a^2)^3(w+b^2)(w+c^2)}}, \\ \boldsymbol{C_b} = \dfrac{2}{\boldsymbol{abc}\displaystyle\int_0^\infty \mathrm{d}w\Big/\sqrt{(w+a^2)(w+b^2)^3(w+c^2)}}, \\ \boldsymbol{C_b} = \dfrac{2}{\boldsymbol{abc}\displaystyle\int_0^\infty \mathrm{d}w\Big/\sqrt{(w+a^2)(w+b^2)(w+c^2)^3}}. \end{cases} \tag{8}$$

By the fields analogy in the field of the electric current, the vector $\boldsymbol{J}$ of the current density and in the magnetic field the vector $\boldsymbol{B}$ of the magnetic flux density correspond to displacement vector $\boldsymbol{D}$ in the electrostatic field. Evaluating relation between the vectors $\boldsymbol{E}$ and $\boldsymbol{D}$ we can interconnect the components of vector $\boldsymbol{D_p}$ inside the ellipsoid D_{ap}, D_{bp}, and D_{cp} with the vector $\boldsymbol{D_0}$ components D_{a0}, D_{b0}, and D_{c0} before the ellipsoid gets into the field this way:

$$\begin{aligned} D_{ap} &= \varepsilon_p \varepsilon_0 E_{ap} = \frac{\varepsilon_p C_a \varepsilon_f \varepsilon_0 E_{a0}}{(C_a - 1)\varepsilon_f + \varepsilon_p} = \frac{\varepsilon_p C_a}{(C_a - 1)\varepsilon_f + \varepsilon_p} D_{a0} = \frac{C_a}{(C_a - 1)^{\varepsilon_f}/_{\varepsilon_p} + 1} D_{a0} \\ &= A_a^{\varepsilon} D_{a0}, \\ D_{bp} &= \frac{C_b}{(C_b - 1)^{\varepsilon_f}/_{\varepsilon_p} + 1} D_{bp} = A_b^{\varepsilon} D_{b0}, \\ D_{cp} &= \frac{C_c}{(C_c - 1)^{\varepsilon_f}/_{\varepsilon_p} + 1} D_{c0} = A_c^{\varepsilon} D_{c0}, \end{aligned} \tag{9}$$

where ε_0 is absolute permittivity of vacuum, and $A_a^{\varepsilon}, A_b^{\varepsilon}, A_c^{\varepsilon}$ are the components of matrix factor A^{ε} which relates vectors $\boldsymbol{D_p}$ and $\boldsymbol{D_0}$. There is convenient to describe the components of A^{ε} this way:

$$A_a^{\varepsilon} = 1 + (C_a - 1)\kappa_a^{\varepsilon}, A_b^{\varepsilon} = 1 + (C_b - 1)\kappa_b^{\varepsilon}, A_c^{\varepsilon} = 1 + (C_c - 1)\kappa_c^{\varepsilon}, \tag{10}$$

where $\kappa_a^{\gamma}, \kappa_b^{\gamma}, \kappa_c^{\gamma}$ are the factors of the electrostatic properties:

$$\kappa_a^{\gamma} = \frac{1 - (\varepsilon_f/\varepsilon_p)}{1 + (C_a - 1)(\varepsilon_f/\varepsilon_p)}, \kappa_b^{\gamma} = \frac{1 - (\varepsilon_f/\varepsilon_p)}{1 + (C_b - 1)(\varepsilon_f/\varepsilon_p)}, \kappa_c^{\gamma} = \frac{1 - (\varepsilon_f/\varepsilon_p)}{1 + (C_c - 1)(\varepsilon_f/\varepsilon_p)}. \tag{11}$$

The relation between the displacement vectors $\boldsymbol{D_p}$ inside the ellipsoid and $\boldsymbol{D_0}$ outside the ellipsoid is linear. In the global coordinate system, it is

$$\left[D_{xp}, D_{yp}, D_{zp}\right]^T = [h][A^{\varepsilon}][h]^T \left[D_{x0}, D_{y0}, D_{z0}\right]^T. \tag{12}$$

If the local coordinate system is turned about the x axis of global system by an angle ψ, about the y axis by an angle ν, and about the z axis by an angle φ, the elements of the matrix $[h]$ in the expression (12) is [14]

$$\begin{cases} h_{11} = \cos\nu\cos\varphi, h_{12} = -\sin\varphi\cos\nu, \quad h_{13} = \sin\nu, \\ h_{21} = \cos\psi\sin\varphi + \sin\psi\sin\nu\cos\varphi, h_{22} = \cos\psi\cos\varphi - \sin\psi\sin\nu\sin\varphi, h_{23} = -\sin\psi\cos\varphi, \\ h_{31} = \sin\psi\sin\varphi - \cos\psi\sin\nu\cos\varphi, h_{32} = \sin\psi\cos\varphi + \cos\psi\sin\nu\sin\varphi, h_{33} = \cos\psi\cos\nu. \end{cases} \tag{13}$$

By fields analogy the relations in the global coordinate system among the components of the virtual current and magnetic flux densities inside the particle J_{xp}, J_{yp}, J_{zp}, B_{xp}, B_{yp}, and B_{zp} and the suitable components before the particle get into fluid J_{x0}, J_{y0}, J_{z0}, B_{x0}, B_{y0}, and B_{z0}, we can express analogically to Eq. (12):

$$\left[J_{xp}, J_{yp}, J_{zp}\right]^T = [h][A^{\gamma}][h]^T \left[J_{x0}, J_{y0}, J_{z0}\right]^T, \tag{14}$$

$$\left[B_{xp}, B_{yp}, B_{zp}\right]^T = [h][A^{\mu}][h]^T \left[B_{x0}, B_{y0}, B_{z0}\right]^T \tag{15}$$

In Eq. (14), $[A^\gamma]$ is a matrix factor of the electric properties of the admixtures. We express $[A^\gamma]$ supposing that the axes of the ellipsoidal particle are oriented by the axes of the local coordinate system but this orientation can be any:

$$[A^\gamma] = \begin{bmatrix} A_i^\gamma & 0 & 0 \\ 0 & A_j^\gamma & 0 \\ 0 & 0 & A_k^\gamma \end{bmatrix}, \tag{16}$$

where

$$A_i^\gamma = 1 + (C_i - 1)\kappa_i^\gamma, A_j^\gamma = 1 + (C_j - 1)\kappa_j^\gamma, A_k^\gamma = 1 + (C_k - 1)\kappa_k^\gamma, \tag{17}$$

$$\begin{gathered} \kappa_i^\gamma = \frac{1 - \left(\gamma_f/\gamma_p\right)}{1 + (C_i - 1)\left(\gamma_f/\gamma_p\right)}, \\ \kappa_j^\gamma = \frac{1 - \left(\gamma_f/\gamma_p\right)}{1 + (C_j - 1)\left(\gamma_f/\gamma_p\right)}, \kappa_k^\gamma = \frac{1 - \left(\gamma_f/\gamma_p\right)}{1 + (C_k - 1)\left(\gamma_f/\gamma_p\right)}. \end{gathered} \tag{18}$$

In Eqs. (16)–(18), $i = a$ or b or $c; j = a$ or b or c; $k = a$ or b or c, $i \neq j \neq k$. Electric properties of the admixtures and the fluid in case of the electric current are expressed by γ_p and γ_f–electric conductivities of particles and fluid, accordingly. The κ_i^γ, κ_j^γ, and κ_k^γ are the factors of the electric conductivity.

$[A^\mu]$ in Eq. (15) is a matrix factor of the magnetic properties of the admixtures. When the ellipsoid axes are oriented by the axes of the local coordinate system randomly, it can be expressed as follows:

$$[A^\mu] = \begin{bmatrix} A_i^\mu & 0 & 0 \\ 0 & A_j^\mu & 0 \\ 0 & 0 & A_k^\mu \end{bmatrix}, \tag{19}$$

$$A_i^\mu = 1 + (C_i - 1)\kappa_\mathrm{i}^\mu, A_j^\mu = 1 + (C_j - 1)\kappa_\mathrm{j}^\mu, A_k^\mu = 1 + (C_k - 1)\kappa_\mathrm{k}^\mu, \tag{20}$$

where

$$\kappa_i^\mu = \frac{1 - \left(1/\mu_p\right)}{1 + (C_i - 1)\left(1/\mu_p\right)}, \kappa_j^\mu = \frac{1 - \left(1/\mu_p\right)}{1 + (C_j - 1)\left(1/\mu_p\right)}, \kappa_k^\mu = \frac{1 - \left(1/\mu_p\right)}{1 + (C_k - 1)\left(1/\mu_p\right)}, \tag{21}$$

where μ_p is the permeability of particles, and κ_i^μ, κ_j^μ, and κ_k^μ are the factors of the permeability.

Relating the biggest semi-axis a of the ellipsoid with any of the axes of the local coordinate system and the other two semi-axes b and c with any of the remaining axes, we obtain all six possible positions of the ellipsoid in the local coordinate system. In Eqs. (10)–(15), it can be six possible combinations of i, j, k:

$$\begin{cases} i = a, \; j = b, \; k = c, \\ i = a, \; j = c, \; k = b, \\ i = b, \; j = a, \; k = c, \\ i = b, \; j = c, \; k = a, \\ i = c, \; j = a, \; k = b, \\ i = c, \; j = b, \; k = a. \end{cases} \tag{22}$$

By Eqs. (13), (16), and (19), we can express the components J_{xp}, J_{yp}, B_{xp}, and B_{yp}:

$$J_{xp} = \left(h_{11}^2 A_i^\gamma + h_{12}^2 A_j^\gamma + h_{13}^2 A_k^\gamma\right) J_{x0} + \left(h_{11}h_{21}A_i^\gamma + h_{12}h_{22}A_j^\gamma + h_{13}h_{23}A_k^\gamma\right) J_{y0}, \tag{23}$$

$$J_{yp} = \left(h_{11}h_{21}A_i^\gamma + h_{12}h_{22}A_j^\gamma + h_{13}h_{23}A_k^\gamma\right) J_{x0} + \left(h_{21}^2 A_i^\gamma + h_{22}^2 A_j^\gamma + h_{23}^2 A_i^\gamma\right) J_{y0} \tag{24}$$

$$B_{xp} = \left(h_{11}^2 A_i^\mu + h_{12}^2 A_j^\mu + h_{13}^2 A_k^\mu\right) B_{x0} + \left(h_{11}h_{21}A_i^\mu + h_{12}h_{22}A_j^\mu + h_{13}h_{23}A_k^\mu\right) B_{y0}, \tag{25}$$

$$B_{yp} = \left(h_{11}h_{21}A_i^\mu + h_{12}h_{22}A_j^\mu + h_{13}h_{23}A_k^\mu\right) B_{x0} + \left(h_{21}^2 A_i^\mu + h_{22}^2 A_j^\mu + h_{23}^2 A_k^\mu\right) B_{y0} \tag{26}$$

We will calculate the average values of the elements of Eq. (6) in two stages. In the first stage, we suppose that the local coordinate axes turned with respect to the global coordinate axes by angles φ, ψ, and ν, and semi-axes of the ellipsoid coincide with the local coordinate axes, but the axes of the local coordinate system and the ellipsoid semi-axes can be any of six possible combinations Eq. (22). The factors $K_F{}^{\mathrm{I}}$ and $K_F{}^{\mathrm{II}}$ we obtain substituting in Eq. (6) J_{xp}, J_{yp}, B_{xp}, and B_{yp} expressed by Eqs. (23)–(26) for any of *i, j, k* combinations Eq. (16) and evaluating that the products of the collinear components $J_x B_x$, $J_y B_y$, $A_i^\gamma A_i^\mu$, $A_j^\gamma A_j^\mu$, $A_k^\gamma A_k^\mu$ are equal to zero. By this we can write

$$\begin{aligned} K_F(\varphi, \nu, \psi) &= \overline{K_F^I(\varphi, \nu, \psi)} = \overline{K_F^{II}(\varphi, \nu, \psi)} \\ &= \overline{H(\varphi, \nu, \psi)} \left(A_a^\gamma A_b^\mu + A_a^\gamma A_c^\mu + A_b^\gamma A_a^\mu + A_b^\gamma A_c^\mu + A_c^\gamma A_a^\mu + A_c^\gamma A_b^\mu\right) \cdot \frac{1}{6} \\ &= \overline{H(\varphi, \nu, \psi)} \cdot \overline{A^\gamma A^\mu}, \end{aligned} \tag{27}$$

where

$$\begin{aligned} \overline{H(\varphi, \nu, \psi)} &= (h_{11}h_{22} - h_{21}h_{12})^2 + (h_{11}h_{23} - h_{21}h_{13})^2 + (h_{12}h_{23} - h_{22}h_{13})^2 \\ &= \Big[h_{11}{}^2(h_{22}{}^2 + h_{23}{}^2) + h_{12}{}^2(h_{21}{}^2 + h_{23}{}^2) + h_{13}{}^2\left(h_{21}{}^2 + h_{22}{}^2\right) \\ &\quad - 2\left(h_{11}h_{22}h_{12}h_{21} + h_{11}h_{23}h_{13}h_{21} + h_{12}h_{23}h_{13}h_{22}\right)\Big]. \end{aligned} \tag{28}$$

We calculate the factors A_a^γ, A_b^γ, A_c^γ, A_a^μ, A_b^μ, and A_c^μ by Eqs. (16)–(18) and Eqs. (19)–(21) substituting $i = a, j = b$ and $k = c$. For the spinning ellipsoidal particle, we calculate the mean value $\overline{\overline{H(\varphi, \upsilon, \psi)}}$ of the $\overline{H(\varphi, \upsilon, \psi)}$ when any of the angles φ, ν, ψ vary in the interval $[0, \pi/2]$. Integrating Eq. (28) we obtain

$$\overline{\overline{H(\phi, \nu, \psi)}} = \frac{8}{\pi^3} \int_0^{\pi/2} \int_0^{\pi/2} \int_0^{\pi/2} \overline{H(\phi, \nu, \psi)} \mathrm{d}\psi \mathrm{d}\phi \mathrm{d}\nu = 1. \tag{29}$$

We express the mean values of coefficients $\overline{\overline{K_F^I}}=\overline{\overline{K_F^{II}}}=\overline{\overline{K_F}}$ when the angles φ, ν, and ψ vary in the intervals [0, π/2] by substituting the $\overline{\overline{H(\varphi,\gamma,\psi)}}$ instead of $\overline{H(\varphi,\gamma,\psi)}$ in Eq. (27):

$$\overline{\overline{K_F}} = \overline{\overline{K_F^I}} = \overline{\overline{K_F^{II}}} = \frac{1}{6}\left(A_a^\gamma A_b^\mu + A_a^\gamma A_b^\mu + A_b^\gamma A_a^\mu + A_b^\gamma A_c^\mu + A_c^\gamma A_a^\mu + A_c^\gamma A_b^\mu\right) = \overline{A^\gamma A^\mu}. \quad (30)$$

We can express the component of the error δ_p caused by the signal variation inside the spinning particle using Eqs. (4), (6), and (30):

$$\delta_p = \frac{\overline{U_p} - \overline{U_0}}{\overline{U_0}} = \frac{\left[\left(1/\tau_p\right)\int_{\tau_p}(W_{zp} - W_{z0})d\tau_p\,\middle|\,\tau_p\right]}{\left[\left(1/\tau_a\right)\int_{\tau_a} W_{z0}d\tau_a\,\middle|\,\tau_a\right]} = \frac{(\overline{W}_{zp} - \overline{W}_{z0})\tau_p}{\overline{W}_{z0}\tau_a} = \left(\overline{\overline{K_F}} - 1\right)k = \left(\overline{A^\gamma A^\mu} - 1\right)k, \quad (31)$$

where τ_p is the volume of the particle and

$$k = \tau_p/\tau_a \quad (32)$$

is the volume concentration of admixtures. $\overline{A^\gamma A^\mu}$ is expressed by Eq. (30).

Eq. (30) is the same for any ellipsoidal particle. It can be generalized for all admixture particles. In this case, τ_p is the volume of all admixture particles.

Substituting Eqs. (17) and (20) into Eq. (31), we can express the error arising inside the admixture particles as follows:

$$\begin{aligned}\delta_p = \frac{1}{6}k\Big\{&2\left[(C_a-1)\left(\kappa_a^\gamma+\kappa_a^\mu\right)+(C_b-1)\left(\kappa_b^\gamma+\kappa_b^\mu\right)+(C_c-1)\left(\kappa_c^\gamma+\kappa_c^\mu\right)\right] \\ &+(C_a-1)(C_b-1)\left(\kappa_a^\gamma\kappa_b^\mu+\kappa_b^\gamma\kappa_a^\mu\right)+(C_a-1)(C_c-1)\left(\kappa_a^\gamma\kappa_c^\mu+\kappa_c^\gamma\kappa_a^\mu\right) \\ &+(C_b-1)(C_c-1)\left(\kappa_b^\gamma\kappa_c^\mu+\kappa_c^\gamma\kappa_b^\mu\right)\Big\}.\end{aligned} \quad (33)$$

4. The error component because of virtual current and magnetic field distortion outside particles

When a particle with volume τ_p and properties different from the fluid physical properties gets into the active zone, it distorts the magnetic field and the virtual current in the residual active zone volume τ_a-τ_p. The variation of the electrode signal ΔU_d because of this distortion will be

$$\Delta U_d = \int_{\tau_a - \tau_p} \overline{\Delta W_{z(-\mathrm{p})}}\mathrm{d}\tau, \overline{\Delta W_{z(-\mathrm{p})}} = \overline{W_{z(-\mathrm{p})}} - \overline{W_{z0}}, \quad (34)$$

where $\overline{W_{z(-\mathrm{p})}}$ are the mean value of the weight vector z component in the any point of volume τ_a-τ_p outside the particle.

We use for ΔU_d investigation the local ellipsoidal coordinate system ξ, η, ζ [13]. The link of this system with the local rectangular coordinate system q, r, s is shown in **Figure 2**.

The coordinate ξ = const is the same on the surface of some ellipsoid, the coordinate η = const is the same on the surface of a hyperboloid with axis q, and the coordinate ζ = const is the same on the two parallel surfaces of a hyperboloid with axis s.

Suppose that equation of the ellipsoidal particle surface is $\xi = 0$ in the ellipsoidal coordinate system. To calculate the function $\overline{\Delta W_z}(\xi)$, we use the Green's theorem written as follows:

$$\int_{\tau-\tau_p} (\mathrm{grad}\psi \mathrm{grad}\varphi)\mathrm{d}\tau = \oint_S \psi(\mathrm{grad}\varphi \mathbf{dS}) - \int_{\tau-\tau_p} \psi \mathrm{divgrad}\varphi \mathrm{d}\tau. \tag{35}$$

Let $\mathrm{grad}\varphi = \boldsymbol{e}_\xi \overline{\Delta W_z(\xi)}$ and $\psi = \int h_\xi d\xi$ (h_ξ is Lame coefficient of coordinate ξ in the ellipsoidal coordinate system). Then $\frac{\partial\psi}{\partial\xi} = h_\xi$ and $\mathrm{grad}\psi = \boldsymbol{e}_\xi \frac{1}{h_\xi}\frac{\partial\psi}{\partial\xi} = \boldsymbol{e}_\xi$. Surface S in the second integral of Eq. (35) is composed of the surfaces $\xi = 0$ and $\xi = \xi_L > 0$. We can write $\mathrm{divgrad}\varphi = \mathrm{div}\left(\boldsymbol{e}_\xi \overline{\Delta W_z}\right) = 0$ because in the volume τ_a-τ_p there are not the sources of $\boldsymbol{J}$ and $\boldsymbol{B}$. Therefore, we can express the electrode signal variation ΔU_d due to the virtual current and the magnetic field distortion this way:

$$\Delta U_d = \int_{\tau-\tau_p} \overline{\Delta W_z(\xi)}\mathrm{d}\tau = \oint_S [h_\xi \mathrm{d}\xi]_S \overline{\Delta W_z(\xi)}\mathrm{d}S = I_{\xi=0} + I_{\xi=\xi_L}; \tag{36}$$

where $I_{\xi=0}$ and $I_{\xi=\xi L}$ are the values of the surface integral on the ellipsoidal particle surface and on the surface ξ_L = const, correspondingly.

Let us investigate the particle with $\gamma_\mathrm{p} \to \infty$ and $\mu_\mathrm{p} \to \infty$. On the surface $\xi = 0$ of such particle, it is right equality $\overline{W_{z\infty}} = (\boldsymbol{J} \times \boldsymbol{B})_{\xi=0} = 0$ because both vectors $\boldsymbol{J}$ and $\boldsymbol{B}$ are perpendicular to the surface at any surface point. Its directions coincide, and the

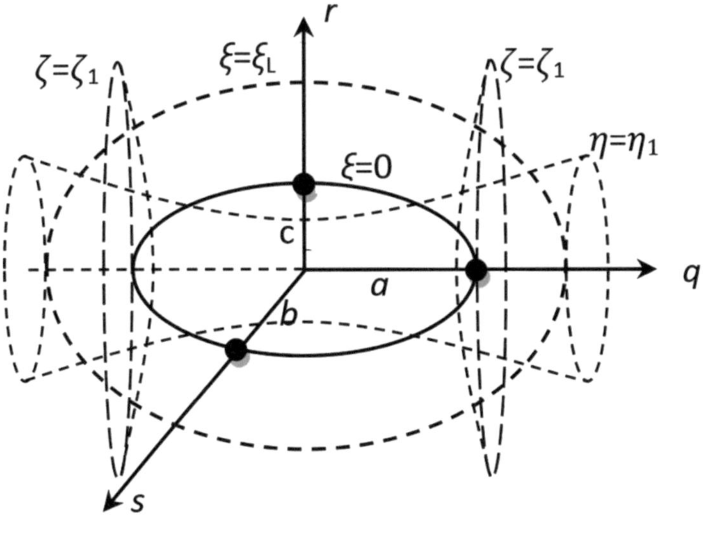

Figure 2.
The rectangular q, r, s and ellipsoidal ξ, η, ζ local coordinate systems.

vector product $[\boldsymbol{J} \times \boldsymbol{B}]$ is equal to zero. This equality is independent on the particle position with respect to the global coordinate system.

By this equality, we can express the mean value of the variation $\overline{\Delta W_{z\infty}}(0)$ of the weight vector z component on the surface of the particle with $\gamma_\mathrm{p} \to \infty$ and $\mu_\mathrm{p} \to \infty$:

$$\overline{\Delta W_{z\infty}}(0) = \overline{W_{z\infty}} - \overline{W}_0 = -\overline{W}_0 = -\left(\overline{J_{x0}B_{y0}} - \overline{J_{y0}B_{x0}}\right). \tag{37}$$

Because $\overline{\Delta W_{z\infty}} = -W_0$ does not depend on the ellipsoidal coordinates ξ, η, and η, the integral $I_{\xi\,=\,0}$ we can write as product of $-W_0$ and integral which expresses the particle volume:

$$\begin{aligned}\Delta U_d = I_{\xi=0} &= \iint_{S_{\xi=0}} \left[\int h_\xi \mathrm{d}\xi\right]_{\xi=0} \overline{\Delta W_{z\infty}}(0)\mathrm{d}S + I_{\xi\mathrm{L}} \\ &= -W_0 \int_{-b^2}^{-c^2}\int_{-a^2}^{-b^2} [h_\xi \mathrm{d}\xi]_{\xi=0}\left(h_\eta h_\varsigma\right)\mathrm{d}\eta \mathrm{d}\varsigma = -W_0\tau_p + I_{\xi\mathrm{L}},\end{aligned} \tag{38}$$

where h_η and h_ζ are Lame coefficients of the coordinates η and ζ in the ellipsoidal coordinate system.

With the increase of the coordinate $\xi_\mathrm{L} > 0$, the shape of ellipsoid $\xi = \xi_L$ is neared to the shape of sphere with the radius $R = \sqrt{\xi + a^2}$ (see [13]) and integral $I_{\xi\,=\,\xi L}$ can be expressed in the spherical coordinates. With R increase, $I_{\xi\mathrm{L}}$ very quickly decreases to zero: $I_{\xi\mathrm{L}} = (-W_0\tau_\mathrm{p})(a^3/R^3) \to 0$. Therefore, outside the particle with $\gamma_\mathrm{p} \to \infty$ and $\mu_\mathrm{p} \to \infty$, the signal variation by Eqs. (30) and (32) is

$$\Delta U_{d\infty} = -\overline{W_0}\tau_p \tag{39}$$

This result can be obtained other way. When $\gamma_\mathrm{p} \to \infty$ and $\mu_\mathrm{p} \to \infty$, the electric resistance for $\boldsymbol{J}$ and the magnetic resistance for $\boldsymbol{B}$ are equal to zero and the densities of virtual current and magnetic flux cannot exchange in the particle volume τ_p. Therefore, the vector product $[\boldsymbol{J} \times \boldsymbol{B}]$ will be equal to zero not only on surface but also in all particle volume. The $\overline{W}_0$will be the same in all volume of clean fluid and in the volume outside a particle with $\gamma_\mathrm{p} \to \infty$ and $\mu_\mathrm{p} \to \infty$. By this, we can write

$$\Delta U_{d\infty} = \int_{\tau_{a-\tau p}} \overline{W}_0 \mathrm{d}\tau - \int_{\tau_a} \overline{W}_0 \mathrm{d}\tau = \overline{W}_0\tau_a - \overline{W}_0\tau_p - \overline{W}_0\tau_a = -\overline{W}_0\tau_p. \tag{40}$$

In [9], it was presented such expression for the signal deviation because the magnetic field and the virtual current distortion $\Delta U_d = -\kappa^\gamma\kappa^\mu W_0\tau_p$. Evaluating that $\kappa^\gamma = 1$ when $\gamma \to \infty$ and $\kappa^\mu = 1$, when $\mu \to \infty$, we can write $\Delta U_d = \kappa^\gamma\kappa^\mu \Delta U_{d\infty}$. This relation is suited for ellipsoidal particles too, but instead the product $\kappa^\gamma\kappa^\mu$ we must use the mean value of this product $\overline{\kappa^\gamma\kappa^\mu}$. The mean value we obtain remembering that relation between the global and the local coordinate axes can be any. Because factors $\overline{\kappa^\gamma}$and $\overline{\kappa^\mu}$are related with perpendicular one to other components $\overline{\Delta J_{x(y)}}$ and $\overline{\Delta B_{y(x)}}$, we obtain the mean value of product $\overline{\kappa^\gamma\kappa^\mu}$ multiplying only the factors $\kappa^\gamma_{i,j,k}$ and $\kappa^\mu_{i,j,k}$ related with the different ellipsoid axes. As a result, we have

$$\overline{\kappa^\gamma\kappa^\mu} = 1/6\left(\kappa^\gamma_a\kappa^\mu_b + \kappa^\gamma_a\kappa^\mu_c + \kappa^\gamma_b\kappa^\mu_a + \kappa^\gamma_b\kappa^\mu_c + \kappa^\gamma_c\kappa^\mu_a + \kappa^\gamma_c\kappa^\mu_b\right). \tag{41}$$

Therefore, in the general case, the mean value of the signal variation $\overline{\Delta U_d}$, due to the virtual current and the magnetic flux distortion outside the particles, will be

$$\overline{\Delta U_d} = -\overline{\kappa^\gamma \kappa^\mu} W_0 \tau_p. \tag{42}$$

We obtain the error δ_d because of the virtual current and the magnetic flux distortion outside the particles as the ratio between signal variation $\overline{\Delta U_d}$ outside particle and signal induced in volume τ_a-τ_p before the particle gets in active zone:

$$\delta_d = \frac{\overline{\Delta U_d}}{\int\limits_{\tau_a - \tau_p} \overline{W_0} d\tau} = -\overline{\kappa^\gamma \kappa^\mu} \frac{\overline{W_0 \tau_p}}{\overline{W_0}(\tau_a - \tau_p)} = -\overline{\kappa^\gamma \kappa^\mu} \frac{k}{1-k} \tag{43}$$

The relation $|\overline{\kappa^\gamma \kappa^\mu}| \leq 1$ is right for any values of γ and μ, so the error because of distortion of the virtual current and the magnetic field practically will not exceed the particle concentration $|\delta_d| \leq k/(1-k)$. For nonmagnetic particles and for particles with $\gamma_p = \gamma_f$ $\delta_d = 0$.

5. The partial error caused by a deviation of mean value of magnetic flux density outside particle

We consider that at the time of measurement, the excitation current of the magnetic field is the same, I_M = const. Therefore, the magnetic field strength H_0 = const will also be the same. If there are no admixtures or they are nonmagnetic, then the magnetic flux density will be expressed by equation $B_0 = \mu_0 H_0$. Let's consider how magnetic flux density B_y will change if a magnetic elliptical particle enters the active zone. If magnetic particles get into the active zone of the electromagnetic flow meter, the mean value of permeability μ_m varies. Expression of μ_m was obtained for the spherical magnetic particles in [9]. For the spinning ellipsoidal particle, we can use this expression too, but factor A_a^μ we must exchange by the its mean value $\overline{A^\mu}$:

$$\overline{\mu_m} = 1 + \frac{\tau_p}{\tau_a} \overline{A^\mu} \left(1 - \frac{1}{\mu_p}\right). \tag{44}$$

In the case of the ellipsoidal particle, the factor $\overline{A^\mu}$ can be expressed by evaluating that the variation of the magnetic flux density mean value does not influence the virtual current. By this, we can write of Eq. (6):

$$\overline{W_{zm}} = J_{x0} \overline{B_{ym}} - J_{y0} \overline{B_{xm}}. \tag{45}$$

We present the mean values $\overline{B_{ym}}$ and $\overline{B_{xm}}$ in two stages as early. At first, we suppose that local coordinate system related with ellipsoidal particle is turned with respect to the global coordinate system by angles φ, ψ, and ν, but the axes of the ellipsoid can be oriented randomly. For that we exchange in Eqs. (19)–(21) $i = a, j = b$, $k = c$ and in expressions (25) and (26), instead $A_i^\mu, A_j^\mu, A_k^\mu$ we use the mean value $\overline{A^\mu}$, expressed as follows:

$$\overline{A}^{\mu} = \tfrac{1}{3}\left(A_a^{\mu} + A_b^{\mu} + A_c^{\mu}\right). \tag{46}$$

We can express of Eqs. (21) and (20):

$$\kappa_a^{\mu} = 1 - \frac{A_a^{\mu}}{\mu_p}, \kappa_b^{\mu} = 1 - \frac{A_b^{\mu}}{\mu_p}, \kappa_c^{\mu} = 1 - \frac{A_c^{\mu}}{\mu_p}. \tag{47}$$

For the spinning particle, the mean value $\overline{\kappa^{\mu}}$ will be

$$\overline{\kappa^{\mu}} = 1 - \frac{\overline{A^{\mu}}}{\mu_p} = \frac{1}{3}\left(\kappa_a^{\mu} + \kappa_b^{\mu} + \kappa_c^{\mu}\right). \tag{48}$$

Evaluating that collinear productions $J_{x0}B_x$ and $J_{y0}B_y$ are equal to zero, we can write of Eq. (26)

$$\begin{aligned}\overline{W}_{zm} &= J_{x0}\overline{B_{ym}} - J_{y0}\overline{B_{xm}} = \overline{\mu_m}\left[\left(h_{21}^2 + h_{22}^2 + h_{23}^2\right)J_{x0}B_{y0} - \left(h_{11}^2 + h_{12}^2 + h_{13}^2\right)J_{y0}B_{x0}\right] \\ &= \overline{\mu_m}\left[K_1^{\mu}J_{x0}B_{y0} - K_2^{\mu}J_{y0}B_{x0}\right].\end{aligned} \tag{49}$$

Coefficients K_1^{μ} and K_2^{μ} by Eq. (13) are.

$$K_1^{\mu} = (\cos\psi\sin\varphi + \sin\psi\cos\varphi\sin\nu)^2 + (\cos\psi\cos\varphi - \sin\psi\sin\varphi\sin\nu)^2 + \sin^2\psi\cos^2\nu, \tag{50}$$

$$K_2^{\mu} = \cos^2\varphi\cos^2\nu + \sin^2\varphi\cos^2\nu + \sin^2\nu. \tag{51}$$

In the second stage, we compute the mean values of coefficients K_1^{μ} and K_2^{μ} integrating the angles φ, ν, and ψ in the intervals $[0, \pi/2]$:

$$\begin{aligned}\overline{K_1^{\mu}} &= \frac{8}{\pi^3}\int_0^{\frac{\pi}{2}}\int_0^{\frac{\pi}{2}}\int_0^{\frac{\pi}{2}}\Big[(\cos\psi\sin\varphi + \sin\psi\cos\varphi\sin\nu)^2 + (\cos\psi\cos\varphi - \sin\psi\sin\varphi\sin\nu)^2 \\ &\quad + \sin^2\psi\cos^2\nu\Big]\mathrm{d}\psi\mathrm{d}\varphi\mathrm{d}\nu \\ &= 1,\end{aligned} \tag{52}$$

$$\overline{K_1^{\mu}} = \frac{8}{\pi^3}\int_0^{\frac{\pi}{2}}\int_0^{\frac{\pi}{2}}\int_0^{\frac{\pi}{2}}\left[\cos^2\varphi\cos^2\nu + \sin^2\varphi\cos^2\nu + \sin^2\nu\right]\mathrm{d}\psi\mathrm{d}\varphi\mathrm{d}\nu = 1. \tag{53}$$

The mean value of the weight vector variation of $\Delta\overline{W_B}$ in active zone outside the admixture particles because of increment of the mean value magnetic flux density is:

$$\Delta\overline{W_B} = \left(\overline{J_{x0}\Delta B_y - J_{y0}\Delta B_x}\right) \tag{54}$$

To find ΔB_y and ΔB_x, we can write

$$\overline{B_m} = \overline{B_f}\frac{\tau_a - \tau_p}{\tau_a} + \overline{A^{\mu}}B_0\frac{\tau_p}{\tau_a}. \tag{55}$$

$\overline{B_m}$ is the mean value of B in the fluid with the admixtures, $\overline{B_f}$ is the mean value of B only in the fluid. Of Eq. (44) we have

$$\overline{B_m} = \overline{\mu_m} B_0 = B_0 + \frac{\tau_p}{\tau_a}\overline{A^\mu}\left(1 - \frac{1}{\mu_p}\right) B_0. \tag{56}$$

Equating the right sides of Eqs. (55) and (56) and expressing B_f we obtain

$$\overline{B_f}\frac{\tau_a - \tau_p}{\tau_a} + \overline{A^\mu} B_0 \frac{\tau_p}{\tau_a} = B_0 + \overline{A^\mu}\frac{\tau_p}{\tau_a} B_0 - \frac{\tau_p}{\tau_a}\frac{\overline{A^\mu}}{\mu_p} B_0 = \overline{A^\mu}\frac{\tau_p}{\tau_a} B_0 + \frac{\tau_a - \tau_p}{\tau_a} B_0 + \frac{\tau_p}{\tau_a}\left(1 - \frac{\overline{A^\mu}}{\mu_p}\right) B_0. \tag{57}$$

Of this equation, we can find $\overline{\Delta B} = \overline{B_f} - \overline{B_0}$ by Eq. (48):

$$\overline{\Delta B} = \left(1 - \frac{\overline{A^\mu}}{\mu_p}\right) B_0 \frac{\tau_p}{\tau_a}\frac{\tau_a}{\tau_a - \tau_p} = \overline{\kappa^\mu}\frac{\tau_p}{\tau_a - \tau_p} B_0. \tag{58}$$

This equation is right for $\overline{\Delta B_y}$ and $\overline{\Delta B_x}$. We express the error for the variation of the magnetic flux density outside a particle δ_B evaluating Eq. (32) this way:

$$\delta_B = \frac{\Delta\overline{W_B}}{W_0} = \frac{\overline{\Delta B}}{B_0} = \frac{\overline{\kappa^\mu} \cdot \tau_p}{\tau_a - \tau_p} = \overline{\kappa^\mu}\frac{\tau_p/\tau_a}{1 - \tau_p/\tau_a} = \overline{\kappa^\mu}\frac{k}{1-k}. \tag{59}$$

Because there is a right the non-equality $\overline{\kappa^\mu} \leq 1$, this error always meets the non-equality $\delta_B \leq \frac{k}{1-k}$.

6. The error because of a variation of the virtual current density mean value

In any section of active zone perpendicular to the axis x, virtual current I_e is the same $I_e = 1\text{A}$. Therefore, if the virtual current density varies inside the admixture particle, it will vary in fluid volume near particle, too. Let δ_J be a relative variation of mean value of the virtual current density x component $\overline{\Delta J_{xf}}$ in the volume of active zone outside the particles τ_a-τ_p: $\delta_J = \overline{\Delta J_{xf}}/\overline{J_{x0}}$. For spinning particles, the values $\overline{J_{y0}}$ and $\overline{\Delta J_{yf}}$ can be related the same equation as $\overline{J_{x0}}$ and $\overline{\Delta J_{xf}}$:$\overline{\Delta J_{yf}} = \delta_J\left|\overline{J_{x0}}\right|$, because the virtual current I_{vy} will be the same in any section perpendicular to the axis y, too: $I_{vy} = 0\text{A}$. Evaluating these relations, we can express the mean value of the variation of z component of the weight vector $\Delta\overline{W_J}$ because of a variation of the virtual current density:

$$\Delta\overline{W_j} = \overline{\Delta J_x B_{y0}} - \overline{\Delta J_y B_{x0}} = \delta_J\left(\overline{J_{x0}B_{y0}} - \overline{J_{x0}B_{x0}}\right) = \delta_J \overline{J_{x0}B_{y0}}. \tag{60}$$

Integrating both sides of this equation in volume τ_a-τ_p, we obtain the electrode signal variation because of the virtual current density mean value variation outside the particles $\Delta\overline{U_J}$:

$$\Delta\overline{U_J} = \int_{\tau_a - \tau_p} \overline{\Delta W_J}\mathrm{d}\tau = \int_{\tau_a - \tau_p} \delta_J W_0 \mathrm{d}\tau. \tag{61}$$

Therefore, δ_j represents a relative error caused by the variation $\Delta\overline{U}_J$.

We want to note one circumstance related to point electrodes. According to the definition given by M.Bevir [7], virtual current flows out of one electrode into another electrode. At and near the electrodes, the current is spread over a very small area, so the density of virtual current in these segments is high. Therefore, it would seem that when the admixture particle is located near the electrodes itself, a very large error is possible. However, we calculate the error by integrating it for the all volume of the active zone (except for the volume of particles), so it is relatively small in this case too.

For any cross section perpendicular to the axis x, we can write

$$J_{xp}\frac{\tau_p}{\tau_a}+J_{xf}\frac{\tau_a-\tau_p}{\tau_a}=J_{xp}k+J_{xf}(1-k)=J_{x0}. \tag{62}$$

Of Eq. (62) we can write

$$J_{xf}=\frac{J_{x0}-J_{xp}\cdot k}{1-k}=\frac{(1-k)-(A^{\gamma}-1)\cdot k}{1-k}J_{x0}$$

and

$$\Delta J_{xf}=J_{xf}-J_{x0}=-\frac{(A^{\gamma}-1)\cdot k}{1-k}J_{x0}. \tag{63}$$

In the case when an ellipsoidal particle is orientated along the x axis by the longest semi-axis, we can express $A^{\gamma}-1=A_a^{\gamma}-1=(C_a-1)\kappa_a^{\gamma}$. For spinning particle, we must use the mean value $\overline{A^{\gamma}}$. Evaluating Eqs. (16)–(18), where $i = a, j = b$ and $k = c$, it can be expressed as:

$$\overline{A^{\gamma}}=\frac{(A_a^{\gamma}+A_b^{\gamma}+A_c^{\gamma})}{3}=1+\frac{(C_a-1)\kappa_a^{\gamma}+(C_b-1)\kappa_b^{\gamma}+(C_c-1)\kappa_a^{\gamma}}{3}. \tag{64}$$

Evaluating Eq. (32) and Eqs. (61)–(64), we can express the partial error which appears because of variation of the mean value of the virtual current outside the particles:

$$\begin{aligned}\delta_J&=\frac{\overline{\Delta J_x}}{J_{x0}}=\frac{\overline{\Delta J_y}}{J_{x0}}=\frac{\Delta \overline{W_J}}{W_0}=-\left(\overline{A^{\gamma}}-1\right)\frac{k}{1-k}\\&=-\frac{1}{3}\left[(C_a-1)\kappa_a^{\gamma}+(C_b-1)\kappa_b^{\gamma}+(C_c-1)\kappa_c^{\gamma}\right]\cdot\frac{k}{1-k}.\end{aligned} \tag{65}$$

7. The common expression of error for magnetic particles and expressions for partial cases

We note the error due to admixtures when electrical and magnetic properties admixtures are different than fluid as admixtures error δ_a. The total error δ_a is sum of the partial error expressions (33), (43), (59), and (65). When $\mu_p > 1$ we can simplify the expressions (43), (59), and (65) as follows:

$$\delta_d^s\cong-\overline{\kappa^{\gamma}\kappa^{\mu}}k,\ \ \delta_B^s\cong\overline{\kappa^{\mu}}k,\ \ \delta_J^s\cong-\frac{1}{3}\left[(C_a-1)\kappa_a^{\gamma}+(C_b-1)\kappa_b^{\gamma}+(C_c-1)\kappa_a^{\gamma}\right]\cdot k. \tag{66}$$

By this, we obtain simplified expression of the total error δ_a^s :

$$\begin{aligned}\delta_a^s &= \delta_p + \delta_d^s + \delta_B^s + \delta_J^s \\ &= \frac{1}{6}\Big\{C_a C_b\left(\kappa_a^\gamma \kappa_b^\mu + \kappa_b^\gamma \kappa_a^\mu\right) + C_b C_c\left(\kappa_b^\gamma \kappa_c^\mu + \kappa_c^\gamma \kappa_b^\mu\right) + C_c C_a\left(\kappa_c^\gamma \kappa_a^\mu + \kappa_a^\gamma \kappa_c^\mu\right) \\ &\quad - C_a\left[\kappa_a^\gamma\left(\kappa_b^\mu + \kappa_c^\mu\right) + \kappa_a^\mu\left(\kappa_b^\gamma + \kappa_c^\gamma - 2\right)\right] - C_b\left[\kappa_b^\gamma\left(\kappa_c^\mu + \kappa_a^\mu\right) + \kappa_b^\mu\left(\kappa_c^\gamma + \kappa_a^\gamma - 2\right)\right] \\ &\quad - C_c\left[\kappa_c^\gamma\left(\kappa_a^\mu + \kappa_b^\mu\right) + \kappa_c^\mu\left(\kappa_a^\gamma + \kappa_b^\gamma - 2\right)\right]\Big\} \cdot k.\end{aligned} \tag{67}$$

Using this equation, we can express errors for some essential cases.

7.1 The expression of maximal value δ_{am} of admixtures error

Maximal value of the admixtures error δ_{am} will be in the case if $\gamma_p \to \infty$, $\mu_p \to \infty$, i.e., for very conductive and very magnetic admixtures. We can write $\kappa_a^\gamma = \kappa_b^\gamma = \kappa_c^\gamma = \kappa_a^\mu = \kappa_b^\mu = \kappa_c^\mu = 1$ in this case, and δ_{am} can be expressed this way:

$$\delta_{am} = 1/3[C_a(C_b - 1) + C_b(C_c - 1) + C_c(C_a - 1)]k. \tag{68}$$

The shape factor in this case is

$$K_{mm} = \delta_{am}/k = 1/3[C_a(C_b - 1) + C_b(C_c - 1) + C_c(C_a - 1)]. \tag{69}$$

7.2 The expression of admixtures error δ_{ae}, if the electric conductivity of magnetic particle is close to the electric conductivity of fluid

If the electric conductivity of magnetic particles with $\mu_p >> 1$ is comparable with the electric conductivity of fluid $\gamma_p \cong \gamma_f$, the following equation is correct: $\kappa_a^\gamma \kappa_a^\gamma \cong \kappa_b^\gamma \kappa_b^\gamma \cong \kappa_c^\gamma \kappa_c^\gamma \cong 0$. The value of the error δ_{ae} in this case will be:

$$\delta_{ae} = 1/3\left(C_a \kappa_a^\mu + C_b \kappa_b^\mu + C_c \kappa_c^\mu\right)k. \tag{70}$$

When $\mu_p \to \infty$, $\kappa_a^\mu = \kappa_b^\mu = \kappa_c^\mu = 1$ and we have maximal error value for this case:

$$\delta_{ae}^{\mu_p \to \infty} = 1/3(C_a + C_b + C_c)k. \tag{71}$$

The shape factor K_{em} is

$$K_{em} = \delta_{ae}^{\mu_p \to \infty}/k = 1/3(C_a + C_b + C_c). \tag{72}$$

7.3 The value of admixtures error δ_{anc} when magnetic admixtures are nonconductive ($\gamma_p = 0$)

In this case.
$\kappa_a^\gamma = -1/(C_a - 1)$, $\kappa_b^\gamma = -1/(C_b - 1)$, $\kappa_c^\gamma = -1/(C_c - 1)\ \kappa_a^\gamma$. Substituting this into Eq. (67), we receive these values:

$$\delta_{anc} = \left\{ \frac{1}{3}(C_a + C_b + C_c) - \frac{(C_a \cdot C_b - C_a - C_b)\kappa_b^\mu + (C_a \cdot C_c - C_a - C_c)\kappa_c^\mu}{6(C_a - 1)} \right. \tag{73}$$
$$- \frac{(C_a \cdot C_b - C_a - C_b)\kappa_a^\mu + (C_b \cdot C_c - C_b - C_c)\kappa_c^\mu}{6(C_b - 1)}$$
$$\left. - \frac{(C_a \cdot C_c - C_a - C_c)\kappa_a^\mu + (C_b \cdot C_c - C_b - C_c)\kappa_b^\mu}{6(C_c - 1)} \right\} k.$$

When the nonconductive particles are very magnetic, we can write $\kappa_a^\mu = \kappa_b^\mu = \kappa_c^\mu = 1$ and we get of Eq. (73)

$$\delta_{anc}^{\mu_p \to \infty} = \frac{1}{3}\left[\frac{C_a}{C_a - 1} + \frac{C_b}{C_b - 1} + \frac{C_c}{C_c - 1}\right]k. \tag{74}$$

The shape factor will be

$$K_{nm} = \frac{\delta_{anc}^{\mu_p \to \infty}}{k} = \frac{1}{3}\left(\frac{C_a}{C_a - 1} + \frac{C_b}{C_b - 1} + \frac{C_c}{C_c - 1}\right). \tag{75}$$

In this case, the shape of particles has no big influence to measurement error. For spherical particles $C_a = C_b = C_c = 3$ and $\delta_{anc}^{\mu_p \to \infty} = 1.5k$. If the semi-axes of ellipsoid are $a = 9$, $b = 3$, $c = 1$, the coefficients are $C_a = 20.4, C_b = 4.44, C_c = 1.39$ and $\delta_{anc}^{\mu_p \to \infty} = 1.97k$. For very elongate ellipsoid with the semi-axes $a = 100$, $b = 10$, $c = 1$ and the coefficients $C_a = 385, C_b = 61.7, C_c = 1.1$ the error is $\delta_{anc}^{\mu_p \to \infty} = 4.33k$.

7.4 The value of admixtures error δ_{anm} when admixtures are nonmagnetic ($\mu_p = 1$)

We note the common error for nonmagnetic particles by δ_{anm}.

In this case, $\kappa_i^\mu \to 0, i = a, b, c$ and simplification Eq. (66) is too rough because in expression (67) $\delta_a^s \to 0$.

We must express total error δ_a using exact expressions of the partial errors. Since we used the multiplier k instead of the multiplier $k/(1-k)$ in the expressions (43), (59), and (65), we need to add to the simplified total error value δ_a^s the sum of partial errors $\delta_d + \delta_B + \delta_J$ multiplied by $k^2/(1-k) = k/(1-k) - k$:

$$\delta_a = \delta_p + \delta_d + \delta_B + \delta_J$$
$$= \delta_a^s + \left\{ -\overline{\kappa^\gamma \kappa^\mu} + \overline{\kappa^\mu} - \frac{1}{3}\left[(C_a - 1)\kappa_a^\gamma + (C_b - 1)\kappa_b^\gamma + (C_c - 1)\kappa_a^\gamma\right] \right\} \cdot \frac{k^2}{1-k}. \tag{76}$$

If $\mu_p = 1$ then $\overline{\kappa^\gamma \kappa^\mu} = 0$, $\overline{\kappa^\mu} = 0$ and we have:

$$\delta_{anm} = -\frac{1}{3}\left[(C_a - 1)\kappa_a^\gamma + (C_b - 1)\kappa_b^\gamma + (C_c - 1)\kappa_a^\gamma\right] \cdot \frac{k^2}{1-k}. \tag{77}$$

7.5 The value of admixtures error δ_{anmc} when admixtures are nonmagnetic ($\mu_p = 1$) and nonconductive ($\gamma_p = 0$)

In this case as in 7.3, we have $\kappa_a^\gamma = -1/(C_a-1), \kappa_b^\gamma = -1/(C_b-1), \kappa_c^\gamma = -1/(C_c-1)$. Evaluating these equalities, we obtain of Eq. (77):

$$\delta_{anmc} = \frac{k}{1-k}^2. \tag{78}$$

Bernier and Brennen in [8] for a flow with air bubbles find that electromagnetic flow meter electrode signal is

$$U_e = U_0/(1-k), \tag{79}$$

where k is the air bubbles concentration and U_0 is the electrode signal for the same flow of fluid without the air bubbles. Murakami et al. [6] experimentally checked the expression (78) and found that it could be successfully used by measuring the liquid and gaseous phase flows separately [12]. A very thorough experimental check of the limits of the application of expression (78) was carried out by J. Cha et al. [7]. They persuaded that if the volume of gas does not exceed 4%, this expression is satisfied precisely. If the flow rate of the suspension does not exceed 30% of the maximum value, then the Eq. (76) is satisfied accurately and by increasing the volume of the gas to 20% of the volume of the suspension. Only with the growth of the suspension flow rate above 30%, the experimental results of the maximum value are slightly different from this expression when the volume of gas in the suspension exceeds 4%.

We will show that Eqs. (78) and (79) express the same result. The flow meter measures some volume V_0 of the flow during some time T. Let the signal of the flow volume V_0 is U_0. The total volume of suspension fluid–air is $V_0(1 + k)$ when the volume air concentration is k. If we measure this volume of homogeneous fluid, the electrode signal must be $V_0(1 + k)$. Therefore, the error δ_{ab} when flow has the air bubbles is

$$\delta_{ab} = 1 + k - 1/_{1-k} = \frac{1-k^2-1}{1-k} = \frac{k}{1-k}^2. \tag{80}$$

This expression coincides with expression (78) for nonconductive and nonmagnetic admixtures.

It is important that in this case the error is not related to the admixture particle shape and depends only on the admixture relative concentration in the second power. The measurement error for nonconductive and nonmagnetic particles can reach 1%, when the volume concentration of particles exceeds 10%.

7.6 The value of admixtures error δ_{acnm} when admixtures are nonmagnetic ($\mu_p = 1$) and very conductive ($\gamma_p \to \infty$)

If $\gamma_p \to \infty \kappa_a^\gamma = \kappa_b^\gamma = \kappa_c^\gamma = 1$. Substituting these equalities to Eq. (77), we have:

$$\delta_{acnm} = -[(1/3)(C_a + C_b + C_c) - 1](k^2/_{1-k}). \tag{81}$$

In this case, measurement error depends on particle shape, but this dependence is actually for very elongate particles, only.

The shape factor we express dividing by k^2:

$$K_{nc} = \delta_{acnm}/_{k^2} \approx -\left[(1/3)(C_a + C_b + C_c) - 1\right]. \tag{82}$$

8. Dependence of measurement error on the shape of particles with different electric and magnetic properties

The measurement uncertainty for conductive and elongate particles can be appreciable in the case of smaller particle concentration. We calculated the diapason of factors K_{mm} and K_{nm} in the cases of very conductive and nonconductive magnetic particles, correspondingly, using program package Matlab. The values C_a, C_b, and C_c were calculated of Eq. (8) and the values of factors K_{mm} and K_{nm} of Eqs. (68) and (72), correspondingly, when $c < b < a$ and the ratios $e = b/a$ and $g = c/b$ were varied in intervals [0.1: 0.95] with the step 0.05. The obtained results are presented in **Figure 3**. We can see that when ratio between longest and shortest semi-axes of ellipsoidal particles is equal to 100, the factor K_{mm} can reach 1434, i.e., the measurement error can exceed the admixtures concentration more than thousand times. Such situation is purely theoretical. Very long and thin particles will be crashed in the turbulent flow or they will not rotate. If they will not rotate, the longest dimension will be directed by flow direction, i.e., by axis z. This dimension has not influence to the measurement error that error will not be great.

We calculated, too, the dependence of the shape factors K_{mm} by Eq. (69), K_{em} by Eq. (72), K_{nm} by Eq. (75), and K_{nc} by Eq. (82). These results are presented in **Tables 1–3** and in **Figures 4–6**.

The particle shape has the great importance to the measurement error when the particles are conductive. We can see that factor K_{mm} quickly increases with the ellipsoid elongation. If the longest and shortest ellipsoid axes lengths ratio a/c is equal to 9, the K_{mm} can be between ≈33 when $b/c = 5$ and ≈54 when $b/c = 1.2$. Therefore, when the concentration of very magnetic and very conductive particles with such shape will be 1%, the measurement error can reach 50%. Fortunately, this situation is few probable because the measurement error to be so great the concentration of long

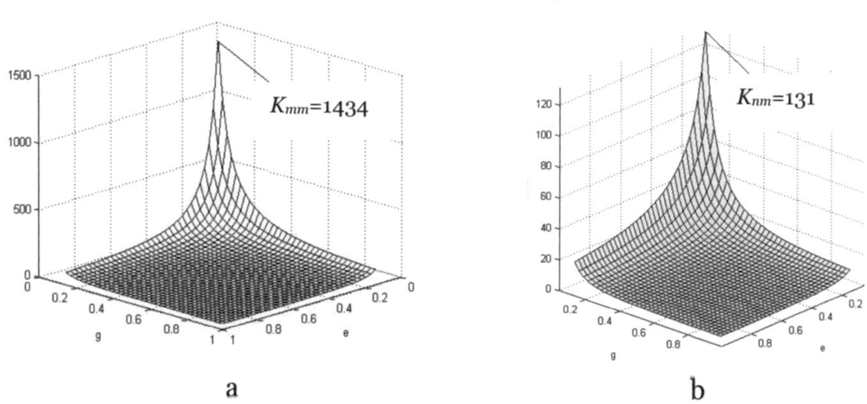

Figure 3.
Dependence of the shape factors K_{mm} (a) and K_{nm} (b) on g and e, when interval of g and e variation is [0.1: 0.95].

a/c	c/b	C_a	C_b	C_c	K_{mm}	K_{em}	K_{nm}	K_{nc}
3	0.9	5.65	5.05	1.61	11.15	4.1	1.7	3.1
3	0.8	5.85	4.58	1.64	10.68	4.62	1.68	3.62
3	0.7	6.1	4.16	1.68	10.22	3.98	1.66	2.98
3	0.6	6.4	3.75	1.74	9.92	3.96	1.63	2.96
3	0.5	6.79	3.34	1.81	9.68	3.98	1.61	2.98
3	0.4	7.29	2.96	1.9	9.63	4.11	1.59	3.11
3	0.3	8.32	2.66	1.95	9.66	4.3	1.59	3.3

Table 1.
Dependence of the shape factors K_{mm}, K_{em}, K_{nm}, and K_{nc} on when ratio between longest and shortest ellipsoid semi-axes a/c is equal to 3.

a/c	c/b	C_c	C_b	C_a	K_{mm}	K_{em}	K_{nm}	K_{nc}
6	0.83	9.83	7.58	1.3	26.1	6.24	2.2	5.24
6	0.67	10.7	6.63	1.34	22.8	6.06	2.03	5.06
6	0.50	12.1	4.62	1.43	20.5	6.05	1.9	5.05
6	0.33	14.9	3.33	1.59	19.6	6	1.73	5.6
6	0.25	17.6	2.68	1.76	20.3	7.35	1.65	6.35

Table 2.
Dependence of the shape factors K_{mm}, K_{em}, K_{nm}, and K_{nc} when ratio between longest and shortest ellipsoid semi-axes a/c is equal to 6.

a/c	c/b	C_a	C_b	Cc	K_{mm}	K_{em}	K_{nm}	K_{nc}
9	0.89	13.8	11.48	1.19	53.9	8.82	2.83	7.82
9	0.78	14.2	9.81	1.21	48.1	8.4	2.63	7.4
9	0.67	14.9	8.41	1.23	43.1	8.2	2.52	7.2
9	0.56	16.0	6.97	1.26	38.8	8.1	2.36	7.1
9	0.44	17.9	5.6	1.31	35.8	8.3	2.17	7.3
9	0.33	20.5	4.36	1.4	32.6	8.76	1.95	7.8
9	0.22	25.8	3.17	1.54	32.6	10.2	1.78	9.2

Table 3.
Dependence of the shape factors K_{mm}, K_{em}, K_{nm}, and K_{nc} when ratio between longest and shortest ellipsoid semi-axes c/a is equal to 9.

very magnetic and very conductive particles must be permanent. With the ratio a/c diminution, the factor K_{mm} quickly decreases. When ratio c/a is not more than 3, the factor K_{mm} < (11–12). It is not more than two times in comparison with K_{mm} = 6 for sphere (see [9]).

When electric conductivity of the magnetic particles is comparable with the fluid conductivity, the dependence on the particle shape is less than for very conductive particle but for the particles with ratio c/a > 9 the error can exceed the concentration 10 times.

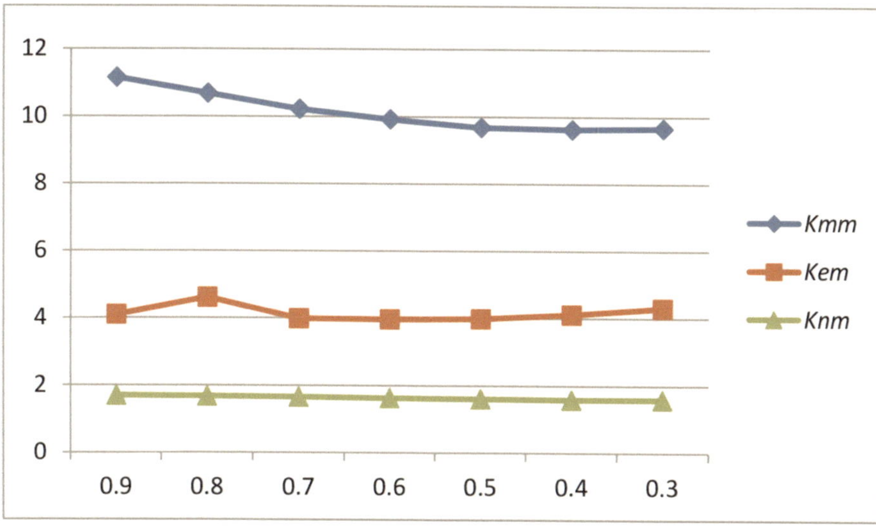

Figure 4.
Dependence of the shape factors K_{mm}, K_{em}, and K_{nm} on c/b when a/c = 3.

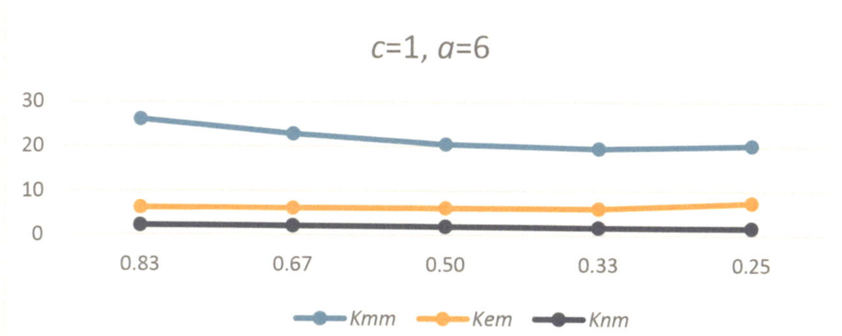

Figure 5.
Dependence of the shape factors K_{mm}, K_{em}, and K_{nm} on c/b when a/c = 6.

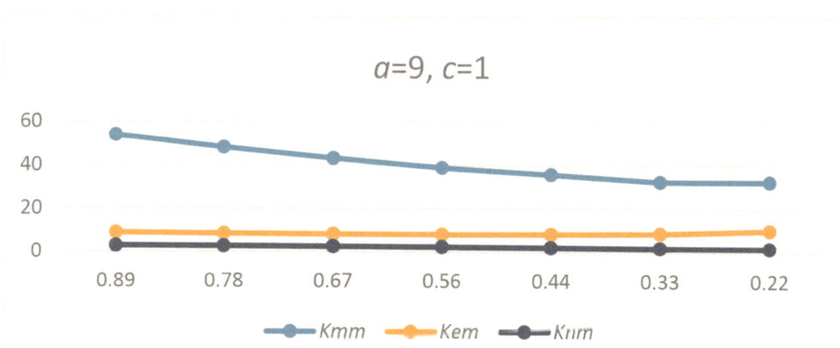

Figure 6.
Dependence of the shape factors K_{mm}, K_{em}, and K_{nm} on c/b when a/c = 9.

For nonconductive magnetic particles, the dependence on the particle shape is yet less than for the conductive particle. For sphere the factor K_{nm} = 1.5, for ellipsoid with a/c = 9 K_{nm} can reach 2.8.

When the admixture particles are nonmagnetic, the measurement error depends on the particles concentration in second power k^2. Therefore, the error will be comparable with concentration of the particles when the factor $K_{nc} \approx 1/k$.

9. Discussion

Very different admixtures can contaminate the fluid flow. The measurement's results will be reliable if we know how the admixtures with the various physical properties influence the measurement results. If we use the electromagnetic flow meters, it is important to know the magnetic and electric features of the admixtures and the shape of the solid admixture particles.

It is convenient to approximate the shape of particles to ellipsoid. By varying lengths of the ellipsoid axes, we can obtain very different spatial profiles: sphere, lamella, cylinder, beam, and other. The ellipsoid has one important advantage for the electric field analysis: when an ellipsoid gets into the homogeneous electric field, there will be the homogeneous field inside the ellipsoid, too. It is a significant property for the analysis of the magnetic field and the virtual current distribution in space with the admixtures. For an effective analysis, the concept of the virtual current proposed by M.Bevir is very important. The virtual current effectively describes the electrical properties of the active zone.

The chosen mathematical tools made it possible to conduct a reliable analysis and find out how various admixtures affect the measurement error.

The performed analysis discovers that the magnetic conductive admixtures are the most dangerous. In this case, the influence of the particle's shape on the measurement error is maximal. For a very long metallic particle with a length ten times exceeding other dimensions, the error has a huge value of δ_{mm} = 1434 k. However, the quantity of very long magnetic particles can't be large. The small particles can't be very long. The long and narrow particles break quickly in the turbulent flow. In addition, very long particles will not rotate in flow. In this case, the maximal dimension will not influence the measurement error.

We can see that the shape of the particle has a lesser influence on the measurement error of the magnetic particles than the particles that are nonconductive. But for very long particles, it increases too.

The measurement error in fluid with the nonmagnetic admixture particles is proportional to the second power of its concentration. Therefore, it is important if the admixture concentration exceeds 5–10%. The error does not depend on the particle's shape when the admixtures are nonmagnetic and nonconductive. The shape of the conductive nonmagnetic particles influences the error but not significantly.

This analysis is performed for the turbulent flow. But we can predicate that in the laminar flow, the measurement error by the same admixtures will be not more than in turbulent. The error can arise for long particles. But the long particles will be orientated along the flow direction in laminar flow, i.e., perpendicular to plane xy. In this case, the longest dimension of the particle will be perpendicular to the components J_x, J_y of the virtual current, and to the components B_x, B_y of the magnetic flux density. Therefore, the longest dimension of particles will not influence the measurement error. The total error will be less than in turbulent flow.

DOI: http://dx.doi.org/10.5772/intechopen.1002509

10. Conclusions

1. The approximation of admixture particles by ellipsoid allows us to investigate how the shape and physical properties of the admixtures influence measurement accuracy.

2. The equation of the measurement error for the general case was derived. The particular expressions were obtained, too, for magnetic nonconductive and very conductive particles, magnetic particles with conductivity close to the fluid conductivity, nonmagnetic nonconductive particles, and nonmagnetic conductive particles.

3. The maximal error arises when admixtures consist of magnetic and very conductive particles with an elongated shape. When the conductivity of the particles decreases, the particle shape influence on the measurement error decreases too.

4. The measurement error is proportional to the second power of the admixture particle concentration when the particles are nonmagnetic.

5. When particles are nonmagnetic and nonconductive, the measurement error is not depended on the particles shape. The expression for the measurement error coincides with expressions received by other researchers.

Author details

R. Račkienė and J.A. Virbalis*
Department of Electric Power Systems, Kaunas University of Technology, Kaunas, Lithuania

*Address all correspondence to: arvydas.virbalis@ktu.lt

References

[1] Shercliff JA. Relation between the velocity profile and the sensitivity of electromagnetic flowmeters. Journal of Applied Physics. 1954;**25**:817-818

[2] Bevir MK. The theory of induced voltage electromagnetic flowmeters. Journal of Fluid Mechanics. 1970;**43**: 577-590

[3] Baker RC. On the concept of virtual current as a means to enhance verification of electromagnetic flowmeters. Measurement Science and Instrumentation. 2011;**10**:105403

[4] Bevir MK. The predicted effects of red blood cells on electromagnetic flow meter sensitivity. Journal of Physics D: Applied Physics. 1971;**4**:387-399

[5] Bernier RN, Brennen CE. Use of the electromagnetic flowmeter in a two-phase flow. International Journal of Multiphase Flow. 1983;**9**(3): 251-257

[6] Murakami M, Maruo K, Yoshiki T. Development of an electromagnetic flowmeter for studying gas-liquid, two-phase flow. International Journal of Chemical Engeneering. 1990;**4**:699-702

[7] Cha J-E, Ahn Y-C, Kim M-H. Flow measurement in an electromagnetic flowmeter in two-phase bubbly and slug flow regimes. Flow Measurement and Instrumentation. 2002;**12**:329-339

[8] Muhamedsalih Y, Lucas GP, Meng YP. Two-phase flow meter for determining water and solids volumetric flow rates in stratified, inclined solids-in-water flows. Flow Measurement and Instrumentation. 2015;**45**:207-217

[9] Virbalis JA. Errors in electromagnetic flow meter with magnetic particles. Flow Measurement and Instrumentation. 2001;**12**(4):275-282

[10] Šimeliūnas R, Virbalis JA. Investigation of field of ellipsoidal shape magnetic particle. Electronics and Electrical Engineering, ISSN 1392-1215. 2002;**nr.** 7(42):72-77

[11] Pakėnas V, Virbalis JA. Influence of non-magnetic amixtures to the signal of electromagnetic flow meter. Electronics and Electrical Engineering, ISSN 1392-1215. 2011;**nr. 10**(116):7-10

[12] Virbalis JA, Račkienė R, Kriuglaitė-Jarašiūnienė, Otas K. The influence of admixtures to the signal of an electromagnetic flow meter. Energies. 2019;**12**:772. DOI: 10.3390/en12050772

[13] Landau LD, Lifshitz EM. Electrodynamics of Continuous Media. Moskow: Nauka; 1982. p. 620

[14] Andre A. Complements de mathematiques. Paris; 1957. p. 708

Chapter 7

Early Advancements in Turbulence-Generated Noise Modelling: A Review

Siddharth Rout

Abstract

Turbulent flows generate a broadband of acoustic noise, which can be extremely important. So, there is need for modelling the generation and propagation of acoustic energy in fluid flows, especially turbulent. This chapter reviews the research work conducted to identify and quantify the noise field generated in turbulent flows. The story starts with the journey of experimental identification and measurement of noise generated from vortices. Various analytical models there were developed, soon after, the popularity of turbulence generated is discussed. The base path-breaking research on quantifying noise generation from conservation laws including Navier–stokes equations is discussed and further used for approximation of acoustic intensity by acoustic analogy with electrostatic quadrupole near-field and far-field. With the development of computational numerical techniques flow field for complex geometries and higher fidelity became possible. The candidates for relevant computational methods are touched and integration with turbulent models is discussed. Finally, a case of simulation of noise generation for turbulent flow over airfoil using acoustic equations and Reynolds-averaged Navier-Stokes (RANS) turbulent model is reviewed.

Keywords: aeroacoustics, hydrodynamic noise, turbulent boundary layers, turbulence theory, acoustic analogy, computational aeroacoustics

1. Introduction

Not just the fans, mechanical components and motors in machines make noise. Noise is also generated from turbulent flows. In fact, these noises are quite common as turbulence is quite common by itself. Starting with milder cases, the sound we hear on the beach is purely due to turbulence. The flowing turbulent fluid through air vents found in buildings and vehicles for heating, ventilation, and air conditioning, commonly called HVAC applications, generates irritating noise affecting normal lifestyle as well as sleep. Denoising in HVAC is a wide spread business in cities as every other building creates a broadband humming noise from their piping systems and chimneys.

In fact, modern cars are designed for streamlined flows across the vehicle to reduce turbulence, subsequently making it more fuel efficient and silent. Opening the windows or sun roof makes the boundary layer flow turbulent and that is the the cause behind the rhythmic noise formation, which mostly experienced in highways once the vehicle crosses a certain speed. In industries, combustion and flame also create turbulence and, hence, generate a broadband and some time-rattling noise from combusters, furnace, and chimney towers. These can be annoying. But are surely annoying in many cases. The early sirens or sirens used in ships and trains were almost completely turbulence generated obtained by the outflows of high-speed turbofans. The Reynolds number has been a well-known parameter to gauge the extremity or severity of turbulence. So, in cases where we encounter flows with larger Reynolds number, like flow through high-speed fans, impellers and propeller, propulsion, jets, ballistics, rockets, aircrafts, etc., it is not just annoying rather unhealthy for physicality and mentality of living beings. Journalists and activists as well as researchers have been raising concerns about noise from helicopters and flights harming residents around helipads and airports [1, 2]. Recently, drones are new additions to the list of generators of loud noise. Researchers are working on reducing such noise as well as psychoacoustic testing for understanding the effect on human psychology [3, 4].

2. Historical development on flow acoustics pre-1950

Acoustics as a branch of physics has been being studied for over two and a half 1000 years. Most of the eminent physicists have studied acoustics. Starting from Aristotle who concluded wave nature and periodic compression and rarefaction of acoustic energy transmission. Galileo and Mersenne concluded vibration to be the cause behind acoustic waves, and they both independently laid the foundational theory of vibration. Soon Newton gave the relationship for wave velocity in solids. Then, a series of development followed with the works by Euler, Lagrange, d'Alembert, and Hemlholtz, who laid the mathematical foundation for acoustics in solid and fluid mediums. First John Tyndall in 1869 brought together the past work with his incremental work on singing flames [5]. It was not clearly understood turbulent flow could also have been a reason along with resonance. Soon after Lord Rayleigh made an extensive study on acoustic behaviour in a wide range of setups and configurations in fluid and solid, whose work is considered monumental till date and has propelled the research on acoustics dramatically [6, 7].

Physicists had been studying various aspects of fluid flows. It was not known that fluid flows could produce sound. Many experiments were being conducted to study wakes and vortices. Luckily, Strouhal in 1878 and Kohlrausch in 1881, quite around the same time, found out about a faint sound originating from vortices, to which the latter described as "reibungstone" [8, 9]. This began to motivate physicists to think about the acoustic aspects of flowing fluids. For quite a long time, numerous works on the new topic started appearing. Mostly experimental and empirical quantification of noise for different cases of fluid flows, like boundary layers, pipe flow, jets, flow past solids, started getting published. Almost all of them used flow parameters like velocity, viscosity, tip speed, non-dimensional quantities like Reynolds number, Strouhal number, etc. [10–13]. Though a wide range of theoretical studies were being conducted, somehow, study on understanding and quantifying the noise generation in fluid flow from first principles was missing.

3. Lighthill's acoustic analogy

Sir James Lighthill, in 1950s, discovered the theoretical connection between fluid flow and acoustics from the conservation laws to derive the wave equation for acoustics [14, 15]. Rayleigh had initiated something in his "Theory of Sound" regarding the origin of sound with which Lighthill could conclude on three ways in which kinetic energy could transform to acoustic energy. They are as follows:

i. By forcing the mass in a fixed region of space to fluctuate, as in a loudspeaker.

ii. By forcing the momentum in a fixed region of space to fluctuate, or, which is the same thing, forcing the rates of mass flux across fixed surfaces to vary; both these occur when a solid object vibrates after being struck.

iii. By forcing the rates of momentum flux across fixed surfaces to vary, as when sound is generated aerodynamically with no motion of solid boundaries.

Based on these three statements, Lighthill could recognise that if flow field generates acoustic energy, then their sources could be derived from Navier-Stokes equation. He creates an experiment assuming a patch of turbulent flowing fluid surrounded by a large domain of surrounding stationary fluid. Let turbulent flow produces noise; however, the noise would transmit to the surrounding fluid at rest. By rearranging and comparing terms in the conservation equations for stationary fluid, the resulting equation could be written as a forced bidirectional wave equation. The forcing term or the source term is what generates noise in flows. Here is the derived wave equation in Einstein's notations,

$$\frac{\partial^2 \rho}{\partial^2 t} - c_o^2 \nabla^2 \rho = \frac{\partial^2 T_{ij}}{\partial x_i \partial x_j} \tag{1}$$

$$T_{ij} = \rho v_i v_j - \sigma_{ij} + \left(p - c_o^2 \rho\right) \delta_{ij} \tag{2}$$

where T is called Lighthill's turbulence stress tensor and has three components or three sources for noise generation. $\rho v_i v_j$ is the convection of momentum fluctuation, σ_{ij} is the viscous stress, and $\left(p - c_o^2 \rho\right) \delta_{ij}$ is the difference in exact p and approximated thermodynamic pressure, $c_o^2 \rho . \rho$ is the density, c_o is the speed of sound, t is the the time dimension, x is the spatial dimension, and v is the velocity. This equation quantifies sound sources from flowing fluid where it takes care of turbulent fluctuations, and viscous dissipation thermodynamics jumps. Now the scientific community is enabled to find acoustic intensity field very accurately from the first principle. This is a remarkable breakthrough in the field of fluid dynamics, which also made Lighthill the father of aeroacoustics. It had been a couple of years with industrial aircrafts in use and both propellers and jet engines produced strong broadband noise. It was a big concern for aerospace researchers and manufacturers, especially during world wars and also after the commercialization of air transport that required more powerful engines to make them faster and carry heavier payloads. In a time of high demand, Lighthill's quantification showed everyone how to use the quantified noise source to find the intensity field right away.

3.1 Lighthill's eighth power law

The major benefit of getting this governing equation, which in fact made it so popular by the name *acoustic analogy*, is the analogy with electrostatics. In a turbulent flow, what vortex stretching creates are acoustic dipoles as the shear creates compression on one side, while rarefaction on the other side. In fact, these dipoles often occur in pairs as vortex stretching creates circulations in couple to balance forces, momentum, and angular momentum. Lighthill could safely assume a turbulent flow field to be distributed acoustic quadrupoles of different powers based on the velocities and vorticity, which typically increases as the size of eddies decreases during vortex stretching due to the conservation of angular momentum. Lighthill could exploit the existing formulations for electrostatic monopoles, dipoles, quadrupoles, etc. That is the base for the eighth power law (**Figure 1**).

A well-established study on electrostatics and electromagnetic radiation theories could be applied to a great extent. A result of that was the Lighthill's Eighth Power Law. Due to the mathematical similarity with electromagnetic radiation theory, the similarity with Stefan-Boltzmann's can also be established where intensity is proportional to the fourth power of temperature. Also, thermodynamic speed is proportional to the root of temperature. So, essentially that gives the intensity of sound radiation is proportional to the eighth power of speed. In aeroacoustics, Lighthill's eighth power law states that the power of the sound created by a turbulent motion, far from the turbulence, is proportional to the eighth power of the characteristic turbulent velocity. We could confer the participating variables, and using dimensional analysis the relation could be found out as

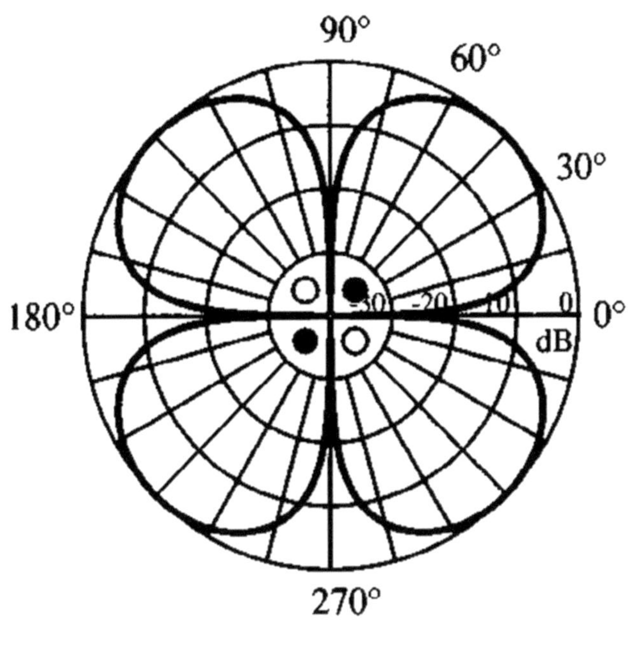

Figure 1.
Theoretical directivity patterns for far-field sound pressure levels radiated from lateral quadrupole, commonly found in vortices [16].

$$W = K\frac{\rho_o}{c_o^5}L^2U^8 \tag{3}$$

where W is the acoustic power in the far-field, K is the proportionality constant (or Lighthill's constant), ρ_o is the uniform fluid density, c_o is the speed of sound, L is the characteristic length scale of the turbulent source, and U is the characteristic velocity scale of the turbulent source. The value of K could be determined experimentally. Depending on a case like hot or cold jet or Mach number range or flow restrictions, the values for K was to be determined experimentally. The eighth power is experimentally verified and found to be accurate for low speed flows, i.e., Mach number is small, $M < 1$. And also, the source has to be compact to apply this law.

3.2 Statistical extension for turbulent jets

After sometime, Lighthill released his statistical work on theories for turbulent cases as an extension to the previous publication [15]. Soon after the theoretical work, acoustic power coefficient (K) and acoustic efficiency (η) were identified from cold jet and hot jet experiments. The experiments as mentioned in Refs. [17–21] were appreciated and used by Lighthill for finding the K and η.

$$K = \frac{W}{\rho_o L^2 U^8 c_o^{-5}} \tag{4}$$

$$\eta = \frac{W}{\frac{1}{2}\rho_o U^3 \frac{1}{4}\pi L^2} \tag{5}$$

An important inference from the latter work was order determination for K and η. The K for sub-sonic jet is roughly in the order of 10^{-4} and η for sub-sonic cold jet is roughly in the order of $10^{-4}M^5$ where M is the orifice Mach number.

4. Sixth power law and later developments

It was understood that once fluid flow becomes chaotic, or turbulent, sharp temporal fluctuation in pressure is often evident due to vortex stretching. Since fluid is a material medium, there is a temporal fluctuation in pressure, and acoustic noise is born. Turbulence happens with eddies of various scales of length and time; hence, an evidence of a broad band of noise is very much logical. With higher Reynolds number, the loudness scales up nonlinearly. These acoustic waves can transmit through walls and can be a real problem, particularly, if it is loud which is the case for missiles, rockets, submarines, and aircraft. Hence, the study and analysis of noise generated from turbulence has been a topic of growing interest.

The vortex theory of aerodynamic sound generation explains how sound is produced by fluid flows through the action of vorticity. This theory is supported by rigorous mathematics and is consistent with Lighthill's theory. At low speeds, the theory predicts sound-power output proportional to the sixth power of the flow velocity with the similarity method and simplification, while at higher speeds, an eighth-power law is observed [22, 23]. The theory has practical applications in engineering, and it is relatively easy to calculate the sound field of a flow based on the motion of the vorticity alone. The theory could be successfully applied to a wide range of flow problems, including aeolian tones, oscillatory flow, spinning-vortex problems, and turbulent shear and jet flows. Almost all the cases in hydrodynamics fall in this regime.

Lighthill's work helped build models for different cases of turbulent jets. Validation of physics using Lighthill's analogy became popular. Yet the most influential research in subsequent years in the field was built on Lighthill's analogy. Its applications on simplified cases like, incompressible jet, inviscid jet etc. [24] Curle-Lighthill [25] Ffowcs-Hawking [26] Analogies for solid boundaries (stationary and moving, respectively) became the base for integral methods in computational aeroacoustics.

5. Computational aeroacoustics

Potential field-based velocity field by linearity assumption is insufficient for complex geometries and inaccurate. Computational models are essential for high-fidelity simulations, especially for reacting flows, multi-physics, multi-phase, complex constraints cases, etc. Also no new pathbreaking theoretical development was seen ahead of quadrupole analogy and acoustic source quantification. Nonlinear problems were tough to solve (shocks, reacting flows), yet were valid for finite approximation as the equations are derived from first principle. These methods boomed after mid 1980s as computational fluid dynamics using finite volume and finite element boomed during the same time.

A few popular computational techniques had linearity assumption that was suitable for fluid at almost rest. Hence, fast and lightweight techniques that were popular in general acoustic simulations like ray tracing, finite difference time domain (FDTD), and boundary element method (BEM) did not prove to be very effective in typical turbulent aeroacoustics. Other fast and lightweight techniques are machine learning based reduced order models, which are catching attention these days. There are linear models like proper orthogonal decomposition (POD) and dynamic mode decomposition (DMD) that are being used for turbulent combustion. Nonlinear models like deep neural networks are also very efficient. However, these machine learning models require a bulky lot of appreciable quality simulated or experimental data to develop the model and training (regression) of the model takes a lot of computational time and resources. So, justified use of machine-learning based reduced order model is a hot debate. Hybridised full-order discretized numerical techniques using finite volume (FV), finite element (FE), spectral methods, etc., with turbulence (with URANS, LES, DES, DNS, etc.) and acoustic models (analogy-based integral methods) have been the best choice till date.

Later decomposition scheme is by Ewert and Schröder [27], an essential technique in hybrid computational aeroacoustics. They describe an approach for simulating acoustic fields in space and time by deriving a family of acoustic perturbation equations (APEs) that predict acoustic sources based on unsteady flow simulation. Their APE formulations are shown to be stable and effective for predicting convection effects in mean flows with vorticity. Different source term formulations have been derived for incompressible and compressible flow solutions, with the vortex sound source being the time derivative of the incompressible pressure or the perturbation Lamb vector. The accuracy of the APE system in predicting convection effects has been tested against solutions of the linearized Euler equations for a monopole in a sheared mean flow and for a spinning vortex pair, showing good agreement. The hybrid approach could be compared with highly resolved unsteady CFD simulation for a laminar flow over a cylinder, and the APE systems yield convincing results for the structure of the pressure contours and decay of pressure with increasing distance from the cylinder.

6. Shen-sørensen aeroacoustic modelling

A numerical algorithm for acoustic noise generation is extended to handle turbulent flows [28]. It comprises of two steps: one for viscous incompressible flow and an other for acoustics of inviscid system. The acoustic part could start at any time during the incompressible computation. The flow is split into Reynolds-averaged component and a component corresponding to the turbulent small-scale fluctuations solved by an eddy viscosity model. The eddy viscosity model called Baldwin-Barth one equation model is used to approximate the noise source.

Case-1: Two-dimensional laminar flow past a NACA 0015 airfoil at an angle of attack equal to 20 degrees at Re = 300 and Mach number 0.2 (**Figures 2, 3**).

Case-2: Two-dimensional turbulent flow past a NACA 0015 airfoil at an angle of attack equal to 20 degrees at $Re = 1.5x10^6$ Mach number 0.2 (**Figures 4, 5**).

Comparing the two cases of flow across an airfoil, it could be noticed that acoustic noise is dominated by the Strouhal frequency and its harmonics. The laminar flow case could be validated with Lighthill acoustic analogy. Regarding the noise level

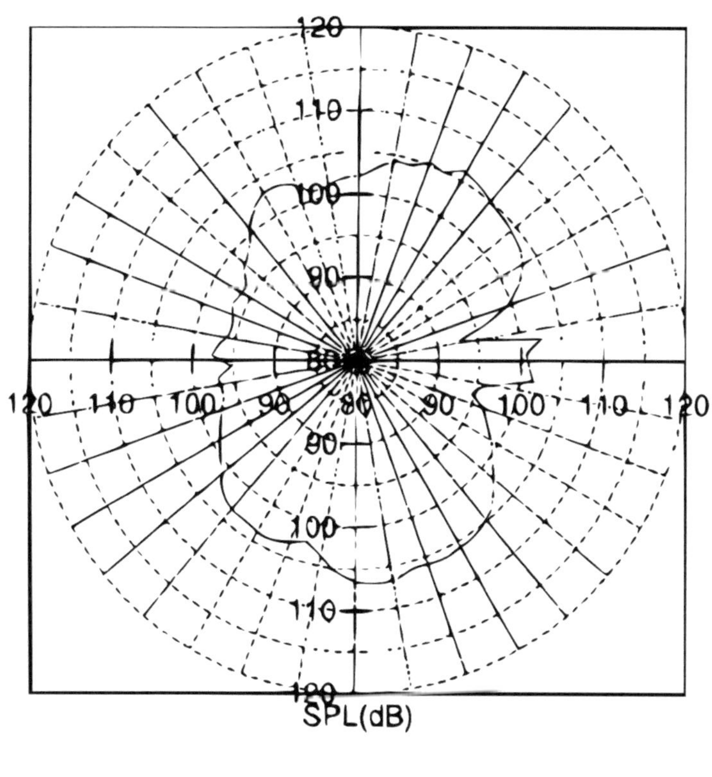

Figure 2.
Directivity pattern of NACA0015 airfoil noise radiation at Re = 300, M = 0.2, and distance = 12.

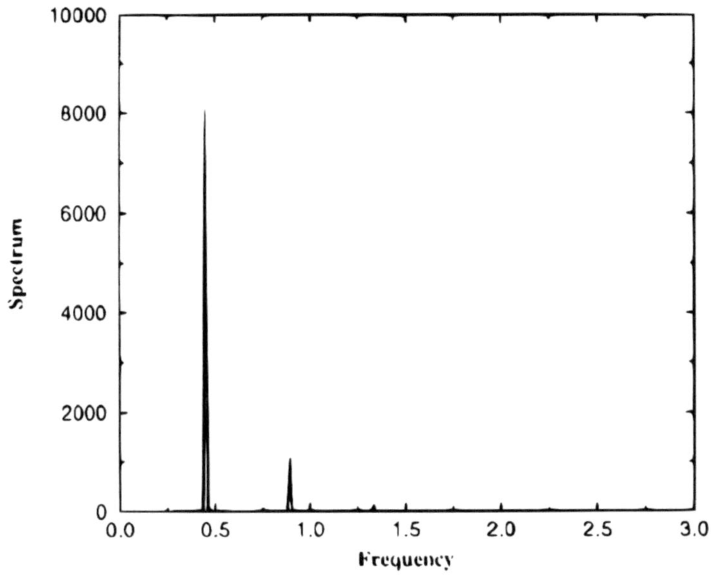

Figure 3.
FFT spectrum of the incompressible vorticity signal for laminar flow past a NACA0015 airfoil Re = 300.

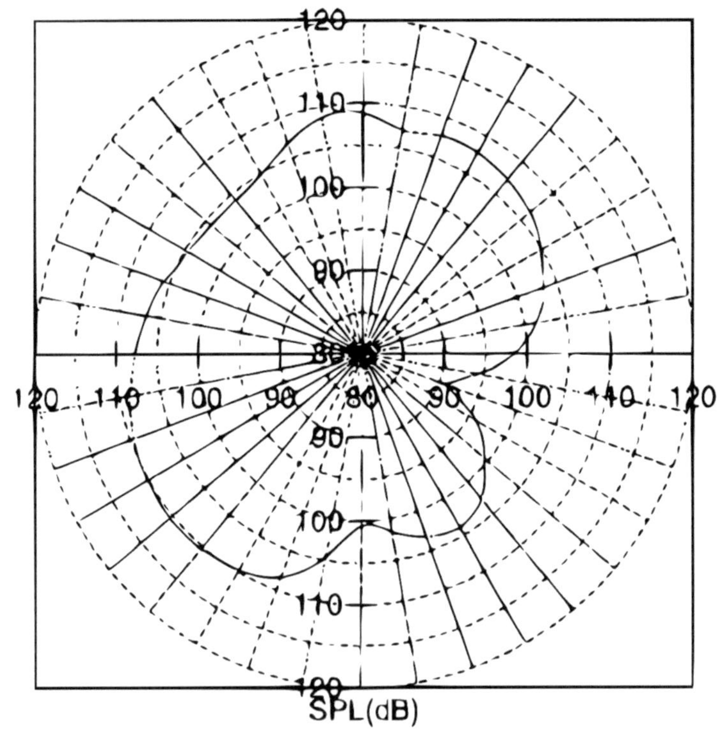

Figure 4.
Directivity pattern of NACA0015 airfoil noise radiation at $Re = 1.5x10^6$, *M = 0.2, and distance = 12.*

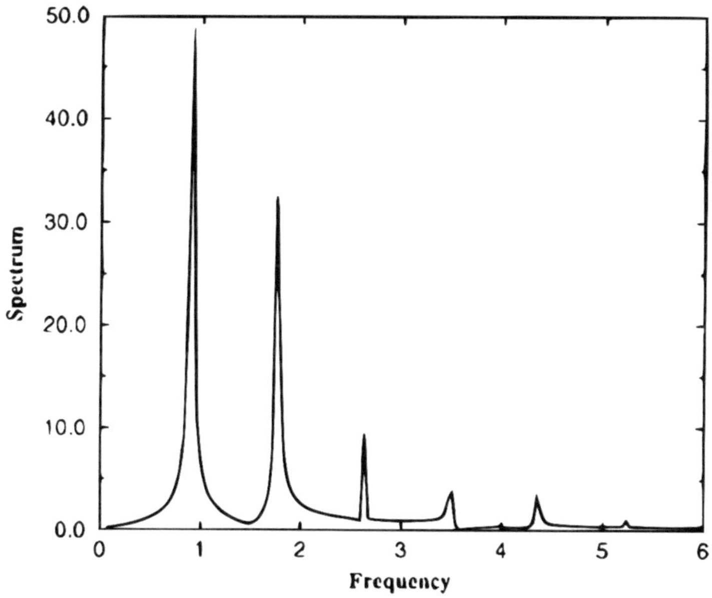

Figure 5.
FFT spectrum of the incompressible vorticity signal for turbulent flow past a NACA0015 airfoil $Re = 1.5x10^6$.

comparison, although turbulent flow is noisier, however both turbulent and laminar cases have the same order of noise level. Only one frequency and its associated higher harmonics were observed for RANS-based model. So, it could be inferred, that needs LES or DNS that captures more frequencies.

7. Conclusions

In conclusion, the study of aerodynamic noise generation has made significant progress in recent years, with a focus on understanding the sources of noise and developing effective computational models for predicting and mitigating noise. Through the analysis of conservation laws, researchers have gained insight into the mechanisms behind both broadband noise and sharp pulses and their harmonics. Lighthill's pioneering work laid the foundation for current computational turbulence acoustic modelling, and ongoing research has led to useful constraints and treatments for modelling complex cases. Despite the annoyance of turbulence-induced noise, these advances bring hope for developing more efficient and quieter technologies in various fields.

Conflict of interest

The authors declare no conflict of interest.

Author details

Siddharth Rout
Department of Mechanical Engineering, University of British Columbia, Vancouver, BC, Canada

*Address all correspondence to: sidrout@mail.ubc.ca

References

[1] Basner M, Clark C, Hansell A, Hileman JI, Janssen S, Shepherd K, et al. Aviation noise impacts: State of the science. Noise & Health. 2017;**19**(87): 41-50

[2] Fort L. Helicopter Noise Making Life a Misery. 2013. Available from: http://www.getreading.co.uk/news/local-news/helicopter-noise-making-life-misery-4190438

[3] Christian AW, Cabell R. Initial investigation into the psychoacoustic properties of small unmanned aerial system noise. In: 23rd AIAA/CEAS Aeroacoustics Conference. Denver, CO, United States: American Institute of Aeronautics and Astronautics; 2017

[4] Gwak DY, Han D, Lee S. Sound quality factors influencing annoyance from hovering UAV. Journal of Sound and Vibration. 2020;**489**:115651

[5] Tyndall J. Sound. New York: Greenwood Press; 1869

[6] Rayleigh JWS. The Theory of Sound. Vol. I. London: Macmillan; 1877

[7] Rayleigh JWS. The Theory of Sound. Vol. II. London: Macmillan; 1878

[8] Stefanini A. Wied. Ann.0. Il Nuovo Cimento. 1895;**54**:264-266

[9] Strouhal V. Ueber eine besondere Art der Tonerregung (On an unusual sort of sound excitation). Annalen der Physik und Chemie. 1878;**5**(10): 216-251

[10] Dryden HL. Recent advances in the mechanics of boundary layer flow. Advances in Applied Mechanics. 1948;**1**: 1-40

[11] Schubauer GB, Skramstad HK. Laminar boundary-layer oscillations and stability of laminar flow. Journal of the Aeronautical Sciences. 1947;**14**(2):69-78

[12] Stowell EZ, Deming AF. Vortex noise from rotating cylindrical rods. The Journal of the Acoustical Society of America. 1936;**7**(3):190-198

[13] Yudin EY. On the vortex sound from rotating rods. Zhurnal Tekhnicheskoi Fiziki. 1944;**14**(9):561

[14] Lighthill MJ. On sound generated aerodynamically. I. General theory. Proceedings of The Royal Society London A. 1952;**211**:564-587

[15] Lighthill MJ. On sound generated aerodynamically. II. Turbulence as a source of sound. Proceedings of The Royal Society London A. 1954;**222**:1-32

[16] Russell DA, Titlow JP, Bemmen Y. Acoustic monopoles, dipoles and quadrupoles: An experiment revisited. American Journal of Physics. 1999;**57**:8

[17] Fitzpatrick HM, Lee R. In: David W, editor. Measurements of Noise Radiated by Subsonic Air Jets Rep. Washington, D.C.: Taylor Model Basin, Wash; 1952. p. 835

[18] Gerrard JH. Investigations of Aerodynamic Sound School [thesis]. UK: The University of Manchester; 1953

[19] Lassiter LW, Hubbard HH. Measurements of Noise Radiated by Subsonic Air Jets Tech. Notes Nat. Adv. Comm. Aero., Washington, D.C.: Wash; 1952. p. 2757

[20] Mawardi K, Dyer I. On noise of aerodynamic origin. Journal of the

Acoustical Society of America. 1953;**25**: 389

[21] Westley R, Lilley GM. An Investigation of the Noise Field from a Small Jet and Methods for its Reduction Rep. UK: Coll. Aero. Cranfleld; 1952. p. 53

[22] Powell A. Theory of vortex sound. The Journal of the Acoustical Society of America. 1964;**36**(1):177-195

[23] Williams JEF. The acoustic analogy—thirty years. IMA Journal of Applied Mathematics. 1984;**32**:113-124

[24] Landau LD, Lifshitz EM. Fluid mechanics. In: Course of Theoretical Physics. 2ed ed. Vol. 6. Oxford, United Kingdom: Butterworth-Heinemann; 1987. p. 75

[25] Curle N. The influence of solid boundaries upon aerodynamic sound. Proceedings of the Royal Society A: Mathematical, Physical and Engineering Sciences. 1955;**231**(1187):505-510

[26] Williams JEF, Hawkings DL. Sound generation by turbulence and surfaces in arbitrary motion. Philosophical Transactions of the Royal Society A: Mathematical, Physical and Engineering Sciences. 1969;**264**(1151):321

[27] Ewert R, Schröder W. Acoustic perturbation equations based on flow decomposition via source filtering. Journal of Computational Physics. 2003; **188**(2):365-398

[28] Shen WZ, Sørensen JN. Aeroacoustic modeling of turbulent airfoil flows. AIAA Journal. 2001;**39**:6